how to build your own spaceship

how to build your own spaceship

the science of personal space travel

piers bizony

Portobello
BOOKS

Published by Portobello Books Ltd 2008

Portobello Books Ltd
Twelve Addison Avenue
London
W11 4QR, UK

Copyright © Piers Bizony 2008

A CIP catalogue record is available
from the British Library

9 8 7 6 5 4 3 2 1

Hardback ISBN 978 1 84627 125 0

Designed and typeset in Optima
by Patty Rennie

Printed in Finland by WS Bookwell ltd

contents

preface

When I was ten years old, my father, who was in the educational publishing business, gave me a scrapbook and some news cuttings and told me that I was to be responsible for scanning all the papers and magazines that came to the house, and snipping out any interesting articles about rockets and astronomy. His company was updating a successful encyclopedia about space. I looked at the first sheath of cuttings he'd handed me: something about a pair of mysterious Soviet capsules that had made a rendezvous in orbit.

I was hooked.

I've maintained ever since that whatever enthusiasm grabs you as a ten year old will never let you go. The cuttings book grew into an elaborate three-volume masterwork. Four decades later, after a varied career in photography, publishing and the media, the space adventure continues to inspire me. But I'm a little bit more grown-up now and some of the themes that intrigue me about the rocket business today are different from the ones that held my attention when I was a child. What happens on the ground is just as fascinating as events in space – and when it comes to the business side, the landscape is changing with incredible speed.

I hope to show you that rocketry is no longer the exclusive playground of giant government agencies. These days it's possible for *you* to get involved. Although it certainly helps if you have access to money, you don't necessarily need frightening amounts. Even if you haven't two pennies to rub together to start with, a little savvy and some good ideas can still get you a foot in the door. The top dozen big players may grab all the headlines for the time being, but thousands of smaller-scale entrepreneurs are involved in new businesses related to space. You could join them.

There's a real buzz, right now, about the coming era of private personal rocket flight, otherwise known as AltSpace or NewSpace. You can see, can't you, from the sliding-together of two words into one that this is a product of the Internet era? Talking of which, if you tap 'space tourist' into the globewide interwub a dazzling array of spacecraft designs will tease you from hundreds of graphically snazzy pages. When you've had your fill of these superficial entertainments and want to know what's real and what's not, that's the time to read *how to build your own spaceship*.

This book is born of my two decades' worth of experience analysing and writing about space flight: not just the hardware and missions, but also the political and corporate wheeling and dealing that makes it all happen – and just as often *prevents* things from happening. When it comes to building our future in orbit and beyond it makes sense to learn from the past. Knowing what I know about the real-life technical workings of the business, I'm not going to try to convince you that building a spaceship is easy or sell you any dreams of a *Star Trek* future looming just on the horizon. That said, I'm convinced that a revolution in space travel is about to be unleashed.

But that's enough about me. Now it's over to you. As Yuri Gagarin shouted when his rocket ship set off on the world's first manned space mission: *Poyekhali!* – Let's go!

piers bizony
november 2007

one   when's the space age coming for the rest of us?

The British songwriter Billy Bragg said recently, 'When I was a kid in the 1960s, I was excited when they told me, "Soon Man will be on the moon." I didn't think they meant just *one* man.' To be fair, a dozen astronauts have explored the lunar surface, but Billy has a point. When's the Space Age coming for the rest of us? Rocket visionaries have often dreamed of sending ordinary people into orbit so that we can colonise the Final Frontier once and for all. That dream has been a long time coming. None of the giant corporations that build today's rockets have been willing to finance the necessary hardware. The most profitable passengers for space have been satellites, not people. Satellites get smaller and cheaper every year and it's just not worth building new and advanced reusable spaceships to launch Comsats the size of desktop computers. So the industry wallows in antiquity. The design of most of today's launch vehicles is pretty much the same as half a century ago, at the very dawn of space flight. Rockets are little more than complicated disposable fireworks. The National Aeronautics and Space Administration (NASA) and its fellow space agencies were once seen as the vanguard of a new technology. Now they may be in danger of becoming the custodians of an old one.

The first stirrings of commercial optimism happened in the 1980s when private businesses were encouraged to pay for NASA Space Shuttle flights in return for access to orbit. The whole thing was a disaster, as two different cultures, each incomprehensible to the other, tried to work out a deal. NASA's complicated rules and regulations put off any but the most ardent bureaucracy junkie, and none of the private companies approached could figure out a way of turning a profit from the space shuttle. Worst of all, the 'reliable, low-cost space truck' promised by NASA turned out to be nothing of the kind. It proved to be one of the least efficient and most dangerous spacecraft ever devised. A generation of space fans brought up on the 'can-do' glory years of the Apollo lunar-landing project watched with dismay as NASA foundered throughout the 1990s. Coming of age (or even middle age), a number of die-hard enthusiasts who'd made it big in high-tech businesses such as computing and the Internet now decided they were rich and powerful enough to revive the space dream without NASA's help.

In 1994 the American Society of Civil Engineers held a conference in Albuquerque, New Mexico, where business leaders began to talk about holidays in orbit. Engineers presented beautiful models of space hotels and passenger-carrying shuttles aimed at a civilian market. Everyone was convinced by the technology. They were equally convinced when financiers told them it couldn't be done, because ordinary folk would baulk at paying silly money for a short spell of weightlessness in a tin can. The pessimists spoke too soon. There are indeed people willing to pay for a taste of space. Perhaps you are one of them?

In 1998 Virginia-based Space Adventures negotiated with Russia's cash-strapped Star City cosmonaut training complex to provide holidaymakers with $10,000 weekends of dressing up in spacesuits and clambering aboard simulator capsules. Customers also signed up for zero-gravity flights aboard 'vomit comets': planes flying a high arc so that they experience a couple of minutes of weightlessness at the top of the arc. This one sorts out the wolves from the chickens. 'Stomach awareness' discourages some people, but if you stay the course you can graduate from flights aboard a converted 707 cargo plane to a trip on a jet fighter.

And that's just the economy end of the Space Adventures menu. On 28 April 2001 the world's costliest holiday got off to a thundering start when American financier Dennis Tito was sealed inside a claustrophobic Soyuz capsule, a design relic dating from the earliest days of Soviet space spectaculars. There was a slight – though still very real – chance that he might not live through the next ten minutes, but he was unconcerned. Tito had invested $20 million of his own cash for this opportunity, booked through Space Adventures. The Russian Space Agency was his airline and the hotel at the end of the flight wasn't some luxury villa by the sea – it was the International Space Station. A year later, South African-born Internet whizz-kid Mark Shuttleworth took the same ride and also lived to tell the tale.

If you've made a fortune in the stock markets or on the Web, $20 million may not be hard to find, but at those prices no one expects too many repeat performances. The future of space tourism will rely on bringing ticket prices down to thousands rather than millions of dollars, so that many reasonably well-off people can afford it, rather than just a handful of multimillionaires. In 2002 the respected market-research company Zogby International surveyed technologically aware, affluent Americans about space tourism. One in five said they would be perfectly happy to pay $100,000 for a suborbital hop, and seven out of every hundred in the super-rich category claimed they would invest millions of dollars in return for a full week in space.

All you need now is a cost-efficient space liner. In May 1996 a group of business leaders held a gala dinner in St Louis, Missouri, to celebrate Charles Lindbergh's historic first solo non-stop flight across the Atlantic in 1927, funded by an earlier generation of patrons from that city. They were competing back then for a $25,000 prize set up by New York hotelier Raymond Orteig. That was a considerable sum of money in 1927. Could a similar prize persuade modern private industry to develop affordable human space flight? And so the famous $10 million Ansari X Prize was born, to reward the first private passenger-carrying ship to reach the edge of space.

Anousheh Ansari couldn't speak English when she emigrated from Iran in 1984 at the age of 16 and came to live in the United States, but she anticipated that living there would help her realise her dream of becoming an astronaut. She enrolled at George Mason University to study electrical engineering. There she met Amir and his brother Hamid, whom she married in 1991. The three of them eventually founded their own telecommunications company. By 1996 Telecom Technologies Inc. was the fifth fastest-growing technology company in Dallas. Anousheh never lost her fascination for the adventure of space flight and she and her company were major sponsors of the Ansari X Prize. 'As a child I looked at the stars and dreamed of being able to travel into space,' she said. 'As an adult, I understand that the only way this dream will become a reality is with the participation of private industry and the creative passion of smart entrepreneurs.' Anousheh made her own flight aboard a Soyuz in 2006 and she continues to promote space tourism. The other significant player in the X Prize is Space Adventures co-founder Peter Diamandis, an energetic entrepreneur whose passion for space is matched by an unprecedented ability to think afresh about the business side of rocketry. He has successfully sold personal adventure 'products' that never existed before: visits to space centres and space stations.

A dozen companies took up the Ansari X Prize challenge. A fascinating array of designs included delta-winged spaceplanes, eerie capsules shaped like seed pods and even a machine with rocket-powered helicopter blades. Some competitors built real hardware, while most were unable to finance much beyond computer-generated concept artworks or unpowered mock-ups. But the seeds of a new industry were sown and a number of companies that didn't make the cut first time around are still in business – older, wiser, less naive about the financial burdens of development, yet undaunted.

One company, Scaled Composites, took a clear lead from the start. Chief designer Burt Rutan's SpaceShipOne scooped the Ansari X Prize on 4 October 2004, immediately earning back half the $20 million invested in the project by Microsoft co-founder Paul Allen. And just as Lindbergh's flight kicked off a transatlantic air industry back in the 1920s, Rutan's team has triggered a similar gold rush at the dawn of the twenty-first century. Just two days after the final qualifying flight, British entrepreneur Richard Branson revealed that his Virgin Group had already pledged another $25 million towards SpaceShipOne's successor, along with appropriate ground facilities in a remote Mojave Desert airport in California. An additional $100 million over the next three years will be spent on a small fleet of 'Virgin Galactic' suborbital space liners, each with six passenger seats, operating from a custom-built space port in New Mexico. For just under $200,000 you can enjoy a precious few minutes of space flight, starting as early as 2010. Branson has been keen on space for many years and his financial projections are encouraging. More than 13,000 would-be space voyagers from around the world have already expressed an interest in paying a deposit to Virgin. They range from anonymous businessmen to glamorous Hollywood stars. Some have paid upfront to guarantee a seat on one of the first flights, but there's nothing stopping you from booking your ticket too. The sign-up procedure is quite simple.

Virgin Galactic does not expect to fly more than one or two missions a week during its first and obviously rather experimental few years. Even so, with bookings worth $1 million per flight, Virgin Galactic could be in profit in just four years. For Branson, space is not just a business opportunity. It's also a personal desire. 'I hope, with the launch of Virgin Galactic and the building of our fleet of spacecraft, that someday children around the world will wonder why we ever thought space travel was just a dream we read about in books or watched, with longing, in Hollywood movies,' he explained, while announcing his venture. 'If we can make space fun, the rest will follow. This is a business that has no limits.' Where Branson leads, you might well follow, or even go further.

The main doubt clouding this bright new horizon is the sheer number of flights that will be needed to make space tourism significantly profitable in the long term – and how those flights will add to the noise and pollution already dumped into our fragile atmosphere by the aviation industry. Other legal and safety aspects are the subject of continuing debate. Even so, SpaceShipOne's success and Branson's plans for Virgin Galactic have boosted the hopes of many companies. Most of them will never get anything off the ground, but among those to watch will be Blue Origin, a Seattle company backed by Amazon founder Jeff Bezos, and the Space Exploration Technologies Corporation (SpaceX), owned by PayPal founder Elon Musk. Just like Paul Allen, Bezos and Musk (and several other players too) can raise the kind of money needed for serious projects. Meanwhile a more conventionally funded company, Constellation Services International – advised in large part by veteran bureaucrats with years of previous insider knowledge of space agencies and governmental procedures – is investing in a wide range of hardware that already exists and is known to work: Russian rocket boosters, crew and cargo capsules, and space-station modules.

And if $200,000 for a seat aboard the Virgin Galactic ship is still too much for you, how about making a day trip into space for little more than the price of a decent car? In Japan, where space tourism is a hugely popular idea, the globe-spanning Kawasaki company has been studying *Kankoh-maru*, a fifty-seat ship shaped like a giant lozenge with a flat base. Kawasaki claims that passengers could pay as little as $50,000 for a ride, which could earn their ship $2.5 million per flight. So now is a good time to pick and choose the companies you might want to invest in. Or maybe you'll decide that their technical approaches are completely wrong and that you have a better plan.

You might want to think higher, faster and longer . . . Hotel entrepreneur Robert Bigelow, multimillionaire founder of the Budget Suites of America hotel chain, is taking the Ansari X Prize idea one step further by offering a $50 million reward for the first company to go all the way into orbit with a reusable passenger ship. That's a much greater challenge than merely dipping your nose above the atmosphere for a few minutes, Virgin Galactic-style, before plunging back to earth, but someone has to do it if space hotels are to become a reality.

Stung into action by all these initiatives, NASA is developing its own award schemes to stimulate yet more private space adventurers. There are various categories with awards running from $250,000 to $50 million for technological innovations; rising to $250 million and more for actual flight hardware. Will it all amount to anything? The first space tourist Dennis Tito certainly wants those less wealthy than him to be able to share his incredible experiences. 'I think one of the things that will keep people interested in space,' he said, 'is seeing other ordinary people go up, from all walks of life. People they can really identify with.'

That's a lovely idea, yet for all the spiritual and scientific arguments in favour of space flight, the cash element causes many people to think twice about its value to society. Perhaps you can find a way of making it a bit less difficult and expensive to accomplish and less of a drain on taxpayers' goodwill? Can you come up with a business scheme that can actually make space profitable? The era of the giant governmental space agency may be coming to a close and the future may now depend on small private companies building a new generation of space technology with private clients funding it and using it. Fantastic opportunities lie ahead. Considerable risks, too. Are you willing to take them? If so, then the sky need no longer be the limit of your reach.

two   that's why they call it 'rocket science'

Let's start with some basic rocket science. Think of catapulting a stone across a field. It flies forwards for a while, then plunges towards the ground as gravity regains the upper hand. Right until the moment of impact the stone continues to travel forwards as well as downwards. Now imagine launching it with much greater force and speed, so that it disappears over the far horizon. This time, the curved surface of the earth continuously drops away beneath the stone once it starts to descend, and it never again comes back to the ground. It becomes a satellite, falling around the planet in a continuous orbit.

In theory a satellite hurled with enough force can orbit at any altitude, so long as buildings, trees or mountains don't get in the way. The lowest practical earth orbit is at around 100 miles altitude. Any lower and the atmosphere slows the satellite down and you fall back to earth.

Any spacecraft must achieve a minimum velocity of 17,500 mph to reach orbit. That's 23 times faster than the speed of sound or 'Mach 23'. The fastest jet aircraft can only reach around Mach 4 – and they rely on the surrounding air not only for lift, but also for burning the fuel inside their engines. The earth's atmosphere thins out almost to nothing at 60 miles altitude. Planes cannot venture to those heights and this is where, according to international convention, the realm of 'space' begins. Rockets are the only sensible means of travelling further. They deliver the necessary power by rapidly converting volatile chemical fuels into pure kinetic energy. According to Isaac Newton, every action has an equal and opposite reaction. You burn fuel and oxygen inside a combustion chamber and hurl the exhaust gases out of the back at extremely high speeds. That's the 'action'. Your rocket is hurled forwards by the back pressure of the exhaust plume pushing against the top of the combustion chamber. That's the 'reaction'. The underlying principle is quite simple, but rocket science isn't called 'rocket science' for nothing. The little details can really tangle you up.

For instance, raw power at the moment of blast off is only one factor in a rocket's design. Overall efficiency during a flight is just as important. Rockets are scaled in terms of *specific impulse*, an equation which measures the thrust delivered over a certain time for a given weight of fuel. A modestly powered engine firing for a long time may get you further than a high-powered engine that burns out in a few moments. On the other hand, a modestly powered engine might not get you so much as an inch off the launch pad, because your rocket is so heavily laden with fuel in those first moments of take-off. The thrust delivered by your system will need to be variable according to circumstances. When you take off from a standing start, your engines need to hurl thunder with utmost fury. Then, once you are in space and already moving along at a good lick, a lower-powered engine firing more gently, but for a longer time, may be more appropriate.

One more subtle design element will be the shape of your engine nozzles, the bell-like protrusions from which the exhaust spurts out. At take-off and during ascent through the dense lower atmosphere, the surrounding air pressure keeps a rocket's exhaust concentrated in a neat plume, maximising the efficiency of the upwards thrust. However, as the atmosphere thins out at high altitudes and you head towards the vacuum of space, the surrounding air pressure drops almost to nothing and the exhaust plume splays out to the sides, losing efficiency. Ideally you want a different-shaped nozzle at this point, longer and thinner to help constrain the plume. In practice it's not easy to design a single type of engine that can play such variable roles. You may end up splitting your rocket into stages, with the first and largest stage powered by the big heavy engines with wide, stubby nozzles, and the upper stages propelled by smaller and lighter engines with narrower and longer nozzles.

The overwhelming complication is the *power-to-weight ratio*. Suppose you build a rocket and then your specific impulse calculation tells you that you need a few more pounds of fuel to sustain the combustion and get the beast all the way into orbit. You change your design by adding an extra few inches to the fuel tank to accommodate the small amount of extra fuel you need. This also means adding an extra few inches of rocket skin, plus pipework to carry the fuel down to the engine. This all adds even more weight to the rocket. The oxidiser tank will also need a bit of adjustment, because you can't burn fuel without the oxidiser. As you factor in these additional weights, the overall mass and bulk of your rocket hardware just keeps rising; so now you need even more fuel to carry the load of the *first* lot of extra fuel, plus the extra weight of rocket skin and pipework. This means adding another few more inches to the length of the fuel tanks, which pushes up the rocket's size and weight again, and so on and so on.

Here's a story from the Apollo moon-landing adventure that puts this vicious circle in perspective. Early in the programme, the astronauts had been looking at various types of shaver, because a few day's growth of stubble on their chins threatened to be uncomfortable when they were wearing their tight-fitting space helmets. The batteries in their Apollo capsule were too precious to be drained by an electrical shaver, so the NASA boffins designed a wind-up mechanical version. But then, what were the astronauts supposed do about the whiskers they trimmed off? In space, they'd float all over the capsule, get in people's eyes, gum up the electronic works. So NASA built a little vacuum cleaner into the razor to suck the bristles away. Well, the whole thing weighed in at about half a pound. That doesn't sound like very much, but it would have taken an additional 50 pounds of fuel inside the giant Saturn rocket at the moment of lift-off to account for getting that little half-pound shaver to the moon and back. NASA decided that the astronauts had better just grow their beards.

This is the main reason why rockets tend to come in separate stages. It's all about losing weight as fast as you can, even while you are climbing into the sky at full thrust. As your huge tanks are drained of fuel at a rate of thousands of gallons per minute, you'll find yourself carrying more and more 'dead weight' in the form of empty tankage, and for no reward, either – just pure penalty in terms of the power-to-weight numbers. The sooner you can dump the depleted tanks, the better. Likewise, when you switch to the smaller upper stage engines for the slow and gentle boost into deeper space, you won't have any further need for the more powerful (and heavier) engines that got you off the launch pad, so these also need to be dumped. Footage of typical rocket flights shows them shedding bits and pieces of kit within a few minutes of leaving the ground, like monstrous metal strippers who can't wait to get down to their smalls. Some rockets drop stages like salamanders pulling free of their tails, and even their torsos, until all that's left is the head. Others shrug off pencil-shaped booster units lying parallel to their sides, until all that remains is the slender spine. Styles may vary, but the song remains the same for all large rockets. It's the 'dropping weight' theme.

rocketing your rocket

So, where does the power come from to drive your rocket? One of the most familiar fuel combinations is kerosene (basically, very high-grade petrol) and oxygen. In principle, an automobile engine burns a similar mixture in its cylinders, although in practice even the noisiest Formula One racing car resembles a rocket about as much as a candle is like a stadium floodlight. Everything happening inside a rocket engine touches upon stupid extremes. At the simple mechanical level, kerosene and oxygen have to be pumped in extremely fast, and at very high pressure, by turbine pumps spinning at upwards of 35,000 rpm. That's twice the spin rate for a jet engine. Building components that don't shatter or distort under these conditions is tough, and expensive.

Then there's the temperature problem. The pressure inside a rocket engine's combustion chamber needs to be as high as the structures will allow, in order to maximise the raw power of the thrust. Temperature raises pressure, and pressure raises temperature. It's pressure you're after and if you could get it from an engine that ran no hotter than a baby's cheek that would save you a great deal of engineering grief. But you can't. Rocket engines run *very* hot indeed. A common protective remedy is to intercept cold fuel, fresh from the tank, and send it around a spiral of tubes coating the outside of the engine to stop it melting under the intense heat of its own workings. Once that snaking path is completed, the fuel – greatly warmed up by now – is sent into the combustion chamber for burning. The pipework for this arrangement, known as 'regenerative cooling', is pretty complicated and very expensive, especially since it only works for a few minutes under such operational extremes: long enough to get a rocket into space, but that's all. The big space agencies don't usually expect liquid-fuelled engines to survive beyond one mission. Because of their raw thrust, they are a familiar power plant for the countless expendable rockets used for launching satellites, planetary probes and capsules. Basically, the expense of building such complex engines for one-time use are justified by the overall scientific or commercial merits of the payload in the rocket's nose.

Another factor is that the 'air' needed for burning the kerosene fuel has to be carried inside the rocket in the form of pure oxygen. The problem is that oxygen takes up an enormous amount of space unless it's compressed, and the best way of doing that is to liquefy it at a super-cold minus 400° Fahrenheit (−240° Celsius) and then store it in an immensely strong tank. You'll have seen liquid oxygen (LOX) tanks being transported on sturdy trucks destined for all kinds of industrial applications. You'll have noticed that those tanks are made from thick, strong steel and consequently tend to be very heavy. For obvious reasons, rocket engineers use LOX tanks that are super-light.

Consequently they are much more difficult to design than the simple steel cylinders favoured by conventional industries.

There's an additional significant trade-off that rocketeers have to make. The thin walls of a booster's LOX tank can put up with the strain while the bulk of the contents is still in liquid form, but as soon as any of it starts to warm up and turn to gas, then the expansion threatens to become a dangerous case of trapped wind, which could, if left unattended, burst the entire tank apart. Moment by moment, the excess gas pressure has to be bled off into the outside world. You can often spot clouds of white smoke streaming down a rocket's sides as it stands against its launch gantry. The vented cold oxygen causes trace amounts of water in the atmosphere to condense around it. It's the water vapour that you can actually see – and you can tell it's not *hot* steam because it spills down the flanks of the rocket, not upwards.

So, a rocket that relies on LOX suffers from a trio of drawbacks. First, the tank is costly and expensive because it has to be light-weight, which means that fabricating it tends to be a job for giant industrial corporations with resources geared up to such a challenge. Second, LOX can only be stored inside a rocket tank for a few hours before launch because eventually it warms up to a gas again, which has to be bled off safely; and third, LOX in its own right is a danger-ously volatile chemical, because, of course, as the supreme oxidiser of all oxidisers, it can burn with *itself*. The slightest inappropriate spark could cause a terrible explosion. Your fuel handling and safety precaution costs go through the roof when you're handling LOX, and likewise your insurance premiums if you're letting private passengers anywhere near a LOX-loaded vehicle. At least the kerosene fuel can't catch fire without the presence of oxygen, so it's not much more hazardous than the petrol in your car. It stays liquid at room temperature, provided you keep it in a sealed container, and behaves itself perfectly well unless mishandled. You can use kerosene in your rocket if you want to, but you'll still be stuck with the LOX problem.

If you are ambitious to lift heavy payloads you'll want a fuel that delivers even more bangs per buck than kerosene. In liquid form, hydrogen is an even lighter fuel, volume for volume, and chemically more energetic. Hence its use in America's mightiest rockets, including the upper stages of the Saturn V lunar boosters for the Apollo missions. The Space Shuttle, for all its flaws, is remarkable for the fact that its hydrogen-fuelled main engines are recovered after each flight. NASA claims that these immensely powerful engines are reusable, but in practice so many of the components need to be stripped out and refurbished after each super-hot flight, they aren't really. A more accurate description would be that they are part-salvageable after what amounts to a controlled inferno.

Hydrogen is difficult to maintain in liquid form and will leak through any microscopic flaw in your tank walls. Just learning how to weld hydrogen tanks reliably was one of the greatest technological breakthroughs that NASA's Apollo engineers ever pulled off. One of the reasons why the Russians lost the race to the moon is that they couldn't handle hydrogen, plus the lift-off weight of their kerosene-fuelled rockets was slightly too high and their thrust slightly too small. Finally, keep in mind that if hydrogen ignites when you don't want it to, you may have the destructive energy of a small atomic bomb to worry about, as NASA learned to its cost when the space shuttle *Challenger* exploded in mid-air in 1986. This stuff is only for serious players.

bang on contact

Another option, if you don't want to bother with super-cold LOX and hydrogen and all that venting, is a system using a fuel and oxidiser combination that can be stored indefinitely at normal pressures and temperatures. Hypergolic chemicals are efficient too. They ignite spontaneously on contact with each other and deliver a pretty good bang for the buck. Of course there's a downside. Hypergolics are among the nastiest and most toxic substances in the rocket business. Did we mention that they can be stored? Well, sort of. They are so corrosive they will play havoc with any part of your rocket (or your people) that they come into contact with when they shouldn't.

On 3 August 1957 the Soviet R-7 missile – or *Semyorka* ('Little Seven') as it was called by the men who worked around it – flew a simulated nuclear strike trajectory. It became a space launcher just two months later, on 4 October, by launching Sputnik. A great international triumph, then, but in missile terms not necessarily delivering quite the military advantage that Russia wanted. *Semyorka* used kerosene and LOX. Who in their right mind wants a nuclear missile that takes three or four hours to prime with LOX before you can launch it? Not the Red Army, for sure. So they commissioned another even more secret missile, the R-16, which in theory could be fuelled and primed several days or even weeks before it was needed with no loss of oxidiser, because its designers had abandoned super-cold LOX and kerosene in favour of nitric acid and hydrazine: hypergolic fuels. In October 1960 the R-16 was hoisted upright for launch at Baikonur Cosmodrome, Russia's ultra-secret equivalent of Cape Kennedy, deep in the deserts of Kazakhstan. And so began the single greatest rocket disaster in history.

The R-16's 'storable' fuels wouldn't store. They were viciously corrosive and leaky as hell, oozing from dozens of pipe joints and tank seams. On 23 October the surrounding launch gantries were crowded with young technicians trying to fix a dozen different problems. As zero-hour approached, the rocket began to drip nitric acid from its base. At this point Field Marshal of Artillery Mitrofan Nedelin, the commander of the R-16 development programme, should have ordered the entire gantry to be evacuated, but he didn't seem to care about the risks. He sent yet more ground staff into the pad area straight away, to see if they could tighten up some valves and stop the leaks and get the rocket up in the air.

Suddenly the rocket exploded, instantly killing everyone on the gantry. With nothing to support it, the upper stage crashed to the ground, spilling fuel and flame. The new tarmac aprons and roadways around the gantry melted in the heat, then caught fire. Ground staff fleeing for their lives were trapped in the viscous tar as it burned all around them. The conflagration spread for thousands of yards, a wave of fire engulfing everything and everyone in its path. More than 190 people were killed, including Nedelin, perched on his chair near the gantry as a surge of blazing chemicals swept towards him.

So maybe you'll want to skip the hypergolic option. Again, this is one for the major players only, and probably best suited to military agencies whose aim is to prevent the Destruction of All of Us by employing rockets that can just kill Quite a Lot of Us.

Now that the realisation has sunk in just how dangerous liquid-fuelled rockets can be, it's time to look at a much simpler propulsion technology, and one so benign (well, sort of) that you're allowed to play with it in your back garden without some government goon knocking on your door with a SWAT team and an arrest warrant.

glorified fireworks

Invented in China some 2000 years ago, the firework has changed little since its origins as a purely ornamental toy. You take a rigid tube of bamboo – or, in our less than all-natural-ingredients Western world, cheap cardboard – and pack the inside with a fine powder made from sulphur, charcoal and saltpetre (potassium nitrate), a common natural mineral. A traditional way of obtaining saltpetre was to filter stale animal manure and urine through a barrel of straw and then let the stinking mess fester for a year or two. Then a water rinse flushed the resulting chemical salts from the straw and the grainy crystalline sediment was filtered through wood ashes and allowed to dry in the sun. China tried to keep the military potential of this stuff secret, but Europeans got the 'gunpowder' bug in the 1300s and have used it ever since as the propellant for – well, guns basically. So that's nice. And by the early 1800s heavy long-range gunpowder rockets were being used in warfare, as the American national anthem *The Star-Spangled Banner*, composed in 1814, so vividly highlights. 'And the rockets' red glare, the bombs bursting in air, / Gave proof through the night that our flag was still there.'

The clever thing about gunpowder is that the chemical mix contains its own oxidiser as well as the fuel component, so no external source of air is necessary for combustion once you've lit the fuse. A gunpowder charge in a cannon, or wedged up tight in the shell casing of a bullet, is loosely packed so that it will ignite all at once. Then the sudden release of explosive gas pushes against the shell and sends it hurtling down the barrel at a suitably lethal rate. The trick with a firework rocket is to *sustain* the explosive reaction over time, so that the rocket continues to accelerate for several seconds after it has left the ground. The gunpowder charge has to be tamed and shaped so that it doesn't simply blow up all at once. Its traditionally powdery consistency has to be melded into something more rigid, a sticky paste held together with oils. There must be no cracks or air pockets in the paste, or else your burn will be uneven.

Your final flourish is to make a neat hole through the exact middle of the paste packing. This is the combustion chamber of your rocket. From microsecond to microsecond, only the innermost wall of the paste tube actually burns away, and the release of energy is therefore gradual rather than instantaneous. Finally, that hole or groove that you've cut into the core is open at one end – the end that all the fire and sparks come shooting out of. As Newton tells us, the energy of the hot gases ejected from the business end creates a reaction against the body of the rocket, driving it forwards, or upwards, depending on which way you point it when you light the fuse. (Actually, this business of pointing rockets in the 'right direction' is what kept firework technology in the slow lane for so many centuries, but we'll talk about that later.)

There are no super-cold liquid fuel tanks, no turbopumps or complicated mazes of pipes and valves; and unlike hypergolics, firework fuels really can be stored for many months or even years at a time, so long as they are kept cool and dry, and, of course, safe from any sources of stray sparks. Fires and floods in firework factories are just so embarrassing, but they happen quite often. The firework is the world's simplest rocket, and for that very reason it has been adapted for many more serious applications. In the early 1940s an amateur chemist turned rocket engineer, Jack Parsons, developed what he called jet-assisted take-off (JATO), a system designed to help heavily bomb-laden American warplanes get off the ground by strapping sophisticated fireworks under their tails. The word 'jet' was chosen to ease the fears of military backers who thought that rockets were Buck Rogers nonsense. NASA's famous Jet Propulsion

Laboratory (JPL) in Pasadena was consolidated around Parsons's work. Apart from rocketry, his other love was the occult. In the late 1940s he was barred from official projects after his dabblings with drugs and louche sexual adventures attracted the FBI's attention. The last straw was when a fellow by the name of L. Ron Hubbard ran off with Parson's wife, plus most of his life savings. Few people remember Parson's name today, yet he co-founded a gigantic new industry devoted to the 'solid rocket motor'. If he'd been a bit more level-headed he could have been a great figure in space flight. Instead he died at just 37 years of age in a chaotic home-based explosion of his own making.

Solid rockets are ideal for military uses, because they are easy to prepare and launch and can be stored for long periods in silos or missile submarines, ready for instant use. NASA uses solid rocket boosters (SRBs) to help lift heavy satellite launchers off the ground. Controversially, two gigantic SRBs strapped to the sides of the space shuttle's fuel tank assist that troubled behemoth into orbit. The snag with SRBs is that they are still just fireworks: once you've lit them, they must burn. They cannot be throttled back, and you can't switch them off if something goes wrong. It was for this reason that veterans from the old Apollo days looked on in dismay when the shuttle designers chose to put SRBs onto a manned vehicle. In January 1986 the space shuttle *Challenger* was destroyed mid-flight because an SRB sprang a small leak of exhaust gas, which hissed like a flamethrower onto the flanks of the shuttle's massive main hydrogen fuel tank . . . You can't necessarily tame a solid rocket's energies after lighting that first spark. Imagine getting into your car and turning the ignition. The engine fires into life and you accelerate rapidly to your cruising speed of 60 mph. All's fine until the time comes to throttle down while you take a sharp corner. No dice – you have to keep going full tilt at 60 mph, because the car cannot be slowed. That's the solid rocket in a nutshell: easy to design, but uncontrollable in a clinch.

Your ideal propulsion unit, then, is much simpler than a liquid engine, but also far more controllable than a solid one. It's a 'hybrid' of the best aspects of both. The hybrid engine is a relatively new technology and the private space sector is leading its development. You might think that the big space agencies would have invented it already if it was any good, but governmental users prefer the more powerful liquid-fuelled rockets suited for heavyweight payloads, despite the risks and costs attendant on handling such volatile systems. Your private project may well have a smaller and lighter ship to launch, perhaps destined only for a brief suborbital trajectory. The hybrid rocket engine, therefore, could be your best solution.

Basically, a hybrid engine consists of a solid-fuelled core through which a liquid or gaseous oxidiser is injected. Nothing can burn unless the oxidiser is whizzing through the core. This ensures that a hybrid can be switched off at any moment if your nerve fails and you decide that now is not, after all, such a good time to punch your ailing ship through the top of the sky. All you have to do is shut the valve and stop the release of oxidiser. As for fine control, that valve can be *partially* opened or shut. The flow of oxidiser through the solid fuel core can be throttled up and down so that the overall burn intensity can be boosted or dampened, depending on how much acceleration you want to achieve at any given moment. Finally there's the ground safety aspect. When your ship is sitting in the hangar there is little prospect of a hybrid engine blowing up. Even if you drench the fuel in leaking oxidiser, nothing will happen without an ignition spark. Using the right combination of materials, you should end up with fuel and oxidisers that cannot burn unless one is ignited in the presence of the other, so you won't risk any kerosene vapour leaks, LOX disasters, or mini-hydrogen-bomb catastrophes.

Nitrous oxide (N_2O or laughing gas) is a safe and effective oxidiser for hybrids, while a good and equally stable fuel core can be made from hydroxyl-terminated polybutadiene (HTPB). That sounds complicated, like you'd need a PhD in chemistry just to pronouce it, let alone refine it. Be ready for a happy surprise as to how cheap and easy HTPB is to source. It's basically powdered tyre rubber. The N_2O is easily stored as a liquid, just like butane camping gas in a lightweight aluminium or tin container. You won't need super-cold tanks to store it; and it has the added advantage of 'self-pressurising', just as butane turns from liquid into a hissing jet of gas as soon as it spurts out of the valve. Your hybrid engine doesn't need complicated turbo pumps or plumbing to push the N_2O oxidiser into the fuel core.

We mustn't run away with the idea that hybrid rocket engines are 100 per cent safe, timid wee creatures just so long as they're not aggravated into an explosion. They – or at least, their pipework and tankage systems – can burst apart with a similar kind of violence to a chemical combustion. If you're standing next to an engine when it malfunctions because of a failed pressure seal rather than a stray spark, you won't be much interested in the difference in the physics as jagged pieces of shrapnel are hurled towards you at bullet-like velocities.

At the extreme end of the pressure scale, NASA's space shuttles are equipped with 24 helium and nitrogen gas tanks that pressurise the shuttle's main propulsion system, orbital manoeuvring engines, and nose-and-tail steering thrusters. Neither helium nor pure nitrogen can burn, let alone explode, yet these systems are fantastically hazardous. The gases are under immense pressure, because their job is to push the volatile rocket propellants into shuttle engines and thrusters at very specific rates required to keep the spaceship on its proper course. Ranging in diameter from 19 to 40 inches, the tanks have lightweight titanium or steel shells wrapped with the same type of fabric used to make bulletproof vests. They hold helium and nitrogen at extremely high pressures of up to 4,600 pounds per square inch. Your little tank of N_2O won't be pumped up to anything like that kind of pressure, but it can still kill you, either by bursting apart because of a pressure-seal failure, or because a small chemical impurity gets into the pure N_2O and gives it something to oxidise at an inappropriate moment. Super-cleanliness and attention to manufacturing detail is essential for all rocket systems, even when they are powered by nothing more obviously frightening than ground-up tyres and laughing gas.

clouds of concern

The more environmentally conscious among you will have spotted that reference to tyre rubber, no doubt with some alarm. A burning tyre emits a nasty black cloud of smoke, so surely this HTPB stuff has to be a no-no? Actually the temperatures and pressures inside a good hybrid engine mean that the burn is relatively clean. The exhaust products take the form of water vapour, carbon dioxide, carbon monoxide and nitrogen. It's not perfect, but still a great deal less environmentally disastrous than the exhaust flares produced by shuttle SRBs, which leave a vast trail of noxious fumes in their wake. The environmental watchdogs will not be unduly concerned at your choice of HTPB while the rest of the world is burning nastier stuff. The space shuttle's main engines, burning a mixture of hydrogen and oxygen, produce nothing more harmful than water vapour as an exhaust product. The SRBs on the flanks of that ship – and many unmanned launch vehicles besides – are a different matter. The combustible core is made from powdered aluminium mixed with ammonium perchlorate, all held together with a sticky plasticised binder. Each booster contains more than a million pounds of this stuff.

During lift-off the three main engine bells under the winged shuttle appear at first glance not to be doing anything. There's no visible smoke, no obvious spurt of fire. This is because the hydrogen/oxygen combo is delivering such a clean burn. Most of the billowing white smoke visible in the first moments of ignition is created when the super-hot downblast from those engines strikes special pools of water under the launch pad. The water is converted instantly to steam and the pad underneath is protected from meltdown. But there's a second and less innocent kind of smoke that puffs up after a few seconds, when the SRBs light up. Noxious clouds deliver an acidic powdery residue into the air. As the shuttle rises into the sky these unsavoury exhaust products are scattered some five miles or more into the surrounding Floridian environment. The residue can irritate the eyes and air passages, and can even spoil the paintwork on your car.

Surrounding the launch complex for many miles is a delicate swampland, a supposedly protected nature reserve rich in wildlife. A few hours after each shuttle lift-off, shallow water lagoons in the swamps become dangerously acidic. Though the pollution fades within a day, the temporary toxicity damages the gills of fish, suffocating as many as a thousand of the hapless creatures per launch. In a paper presented to a meeting of the American Institute of Aeronautics and Astronautics in 1983 on the biological effects of shuttle launches, the US Air Force concluded, 'The affected animals should be considered dedicated in the interests of the mission.' Plenty of people have disagreed with that sentiment, and today NASA is under pressure to rethink its use of propellants, although designs for future space agency launch vehicles still call for the use of SRBs of some kind.

It's not just the shuttle that's to blame. Most of the American space fleet – including the unmanned Titan and Delta families of satellite launchers – rely on similar solid-rocket boosters to provide enough thrust to get them into orbit. The European Space Agency (ESA) also uses solids to give a kick to its Ariane launcher. Russian launches are not very clean either. Expended stages from Soyuz rockets drop onto Kazakhstani farmland, often still containing dregs of unburned kerosene, plus traces of engine lubricants, residues from the explosive charges that help separate the rocket stages, and so on. Each launch may be only a minor inconvenience, but cumulatively, half a century's worth of these things dropping out of the sky has done serious damage to the land, and to the health of the farmers and their families. Wreckage from the larger hypergolic-fuelled Proton cargo rockets is even more poisonous. And let's not forget the ozone layer. Rocket plumes aren't good for that either. The immediate impact of exhaust from a solid-fuel rocket on local stratospheric ozone is not clearly understood, but the assumption among the critics of rocket flight is that the ozone must surely get a hammering. One of the questions that your rocket company will have to address is this: how clean are you?

N_2O hybrid engines have relatively clean exhaust products, second only to the liquid oxygen/hydrogen combination. The oxidiser-to-fuel ratio is typically 6 to 1, so the bulk of the exhaust gas is essentially just super-heated nitrogen from the N_2O. As you no doubt remember from school lessons, the air we breathe is 78 per cent nitrogen. The remaining products of combustion are not so benign, with carbon dioxide and a little carbon monoxide undeniably present in the exhaust trail, but a laughing-gas engine isn't about to destroy the world. Your environmental problems will be largely perceptual rather than real. Can you convince the public that, even though each flight represents just the tiniest drop in a polluted ocean, *all* the flights undertaken by your fleet over the months and years ahead won't have a cumulative effect? To be honest, no you can't, and nor should you try. The rocket that makes zero impact on the atmosphere hasn't been invented yet.

navigating in space

Once your spacecraft has reached orbit, there's no clear magnetic north or south for you to steer by. However, the axis of a freely spinning gyroscope (or gyro) always points in the same direction. Three gyros, angled 90 degrees apart, keep track of a spacecraft's orientation. Meanwhile, sensors known as accelerometers register changes in velocity in any given direction. An inertial navigation system (INS) monitors your spacecraft's position relative to the earth, and especially to the horizon. But the earth itself is constantly moving through space as well, so how do you keep your bearings? A fixed frame of reference is what you need. Before launch the gyros in your craft are spun up with their axes precisely aligned with reference to a particular guide star. No mechanical device is perfect, however, and minor misalignments usually occur during a mission. Spacecraft are usually equipped with an optical star tracker which compensates for unwanted drift between the gyroscopes and the actual direction of the guide star. Canopus, in the southern constellation Carina, is commonly used. If you're heading for orbit or into deep space, you'll need a Canopus Tracker in your ship.

Fine steering control is another matter altogether. Shuttles and capsules use small gas thrusters to control their orientation, or *attitude*. Thrusters are fired in short bursts to *pitch* the nose up or down, to *yaw* the nose left or right, or *roll* the ship around its axis. This is only half the problem. Thrusters firing in the opposite direction must brake your ship at the end of a manoeuvre, otherwise you will continue to tumble. Computer guidance is essential, but you might like to fire thrusters manually during the last few metres of a docking approach, when your target is clearly visible through the cockpit windows. This establishes you as a steely-eyed astronaut and not just a button-pusher. Actually the computers are more reliable, but you won't want to hear that kind of talk if you've been brought up on images of Han Solo hauling at the joystick of his *Millennium Falcon* while twisting and turning through an asteroid field. Seriously. Let the computers do the work. Or at least make sure that a woman handles the controls during tricky dockings. A 1997 US Department of Defense study concluded beyond doubt that women pilots' reaction times and coolness under pressure are better than men's.

Thrusters aren't the only way of controlling attitude. Heavy 'reaction wheel' gyros are not easily dislodged from the plane in which they are spinning. A spacecraft can use actuators to push against the frames that contain each of the gyros, but they will tend to stay in the same orientation, allowing the spacecraft to push against them and change its orientation instead. This system is perfect for robotic space platforms and other satellites which have to function in orbit for years at a stretch without wasting their small reserves of thruster fuel. Reaction gyros are powered by renewable electricity from solar panels. Another advantage of gyros over thrusters is that delicate optical instruments aren't endangered by exhaust gases. Not surprisingly, the Hubble Space Telescope uses gyros rather than thrusters.

In all likelihood, your ultimate aim won't be to stare at the stars, but to make a rendezvous with a space station or an orbiting hotel, so you'll be needing those thrusters. Be warned, though. Orbital mechanics take some getting used to. Suppose you want to dock with a target flying just ahead of you on exactly the same orbit. Accelerating your ship by firing the rockets will simply boost you into a higher orbit, while your speed relative to the ground will decrease. The other spacecraft will appear to hurtle downwards and away from you. On the other hand, firing your breaking thrusters will lower your orbit and *increase* your ground speed. This time your target will appear to shoot overhead and behind you . . . The trick is to begin your rendezvous approach from a slightly higher or lower orbit than your target.

If this doesn't make intuitive sense, then let's look at two extreme examples of different orbits. A 'low earth orbit' at 200 miles altitude will send your ship around the planet once every 90 minutes, at a ground speed of some 17,500 mph. You'll see 16 sunrises and 16 sunsets in a single 24-hour day. From an observer's point of view on the ground, your craft, seen as a tiny speck of light in the night sky, will appear to shoot from west to east, from far horizon to far horizon, in just a few seconds. On the other hand a spacecraft flying a much higher, 'geosynchronous' orbit, some 20,000 miles above the ground, will go around the earth once every 24 hours, at a speed through space of just 6,000 mph, and at a *ground speed* of zero! It will appear to hover over the same spot. So, the higher an orbit you climb up to, the slower your speed. That's why another spacecraft occupying a lower orbit will travel faster than you.

The fuel burned in reaching higher orbits doesn't necessarily translate into speed. The effort is devoted to the long, hard climb upwards in the teeth of gravity's desire to pull you back down. The higher your orbit, the weaker the effect of gravity becomes, and the slower you want to go so as not to escape orbit altogether and accidentally hurtle off into the void. On the other hand, if you do want to break the bonds of gravity altogether, you keep firing your engines while still in low earth orbit, but instead of aiming your nose upwards and draining all your energy climbing up that way, you drive on forwards, at low altitude, translating all your thrust into sheer momentum until you reach a ground speed of 25,000 mph. Then, at last, the earth will surrender its grip and let you go.

round and round

Depending on the mission you want to fly, there are three major classes of orbit that you might want to choose from. The low-earth-orbit (LEO) realm that we've already encountered is probably the one you'll be after. Shuttles and space stations fly near-circular orbits around the earth's equatorial belt at altitudes ranging from 200 miles (where the International Space Station lives) to around 600 miles (the realm of the Hubble Space Telescope), far above the slightest threat of whispy atmospheric disturbance. Most astronaut missions fly in LEO space at 200-odd miles altitude.

Earth's equatorial circumference is 24,860 miles. Since the planet spins on its axis once every 24 hours, a rocket at an equatorial launch site benefits from a free ride eastwards at around a thousand miles per hour even before its engines are ignited. (Launch pad technicians don't have to brace themselves against a thousand mile-per-hour wind, because the air around them is travelling eastwards too.) Major space ports are located as near as possible to the equator to exploit this. However, if you are launching a satellite to map the earth's surface, then an equatorial orbit, as favoured by the astronauts in their heavy ships, isn't going to be of so much use to you, for the simple reason that you don't just want to map the equatorial regions again and again. What about the Antarctic ice sheets and the Norwegian fjords? For these kinds of mission you'll want to launch into a polar orbit. As your satellite flies around the earth, the rotation of the planet around its axis ensures that a slightly different patch of territory comes under the satellite's mapping cameras during each orbit. The downside is that you lose that equatorial free ride from west to east on launch day, and consequently you need more fuel, plus an additional 'kicker' stage under the satellite, to reach a polar orbit.

The third and perhaps most commercially valuable orbit was identified by the British space visionary Arthur C. Clarke back in 1945. A satellite flying a geosynchronous orbit from west to east, on the same plane as the equator, can keep pace with the 24-hour rotation of the earth, hovering motionless over a chosen patch of ground. Three satellites, equally spaced along the orbit, are sufficient to create a seamless global radio link.

If you remember your earlier lessons in orbital mechanics, such a slow ground speed, at which a satellite actually *stays still* relative to the ground, must equate to a very high orbital altitude, right? And in this case the altitude needs to be precise: 22,238 miles. Access to geosynchronous orbit is monitored by international agreement so that the available parking slots directly overhead certain particularly desirable ground locations don't become too crowded. As Clarke predicted, radio and TV satellites love it up here, but it's also a good home for probes that need to monitor the weather in remorseless detail over a specified continent, so that interchangeably bland TV weatherboys and weathergirls of many nations can point to their special-effect blue screens and say, 'Here's the satellite picture for today, and as you can see, we have a band of rain sweeping in from the north . . .'

'Do you read me, HAL?'

Orbital manoeuvres are not the sort of stunt you can pull off just by eye. You're going to need a good computer, linked with a radar antenna that can sense the distance between you and your rendezvous target. If you'd been trying to build your spaceship back in the 1960s, in the earliest years of the Space Age, the computer alone would have cost you many millions of dollars to design and build. Today, any decent university lab should be able to knock something up for you out of 'off-the-shelf' microchips, and for a budget price, too. The only additional problem will be to protect the chips against damage from space radiation. Space-hardened chips should be available. You'll also want 'redundancy' built into your electronics in case any of the circuits fail, so be prepared to double up on your motherboards, CPUs and power supplies, and make sure to run a parallel third system on the ground during every flight. If your pilot experiences an unusual problem with the on-board computers a ground technician should be able to figure out what's gone wrong and advise the pilot which switch to throw. And yes, that advice may well take the form of 'Have you tried switching the computer off and on again?' In the mission control business, they call this 'cycling the hardware' rather than 'wasting the customers' time on a premium-rate phoneline.' HAL 9000 in the movie *2001: A Space Odyssey* (1968) expressed his misgivings about this kind of thing by murdering most of the crew. If you are going to reboot, best not tell your on-board computer.

The unsettling truth is that spacecraft don't really need pilots. NASA understands this perfectly well, and Russia's Progress cargo ferries and the ESA's latest cargo ship, the Automated Transfer Vehicle (ATV), are just that: automated, all the way from orbital insertion to rendezvous and docking with the International Space Station.

In the late 1990s NASA looked at possible replacements for the space shuttle, in which the crew compartment was supposed to be just a slightly specialised cargo container slotted into a completely automatic vehicle, known as VentureStar. Astronauts or hardware were to be treated just the same, as inert payloads accommodated in the belly of the robot, which would blithely fly its mission without responding in any way to what was inside it – it could be people, it could be potatoes. VentureStar couldn't care less. Needless to say, many NASA veterans were uncomfortable with this idea and the illusion of cockpits and control panels will be maintained for future human missions, even though advanced computers will actually be making most of the decisions. As for your own beloved project, you can forget any thoughts of Dan Dare tugging fondly at the control levers of his trusty rocket ship *Anastasia*. Times have changed. Nevertheless, your passengers will want to think that they are in human hands rather than at the mercy of a microchip, so pander to that illusion, perhaps for your own sake as much as theirs.

three runners and riders, ways and means

If you want to get something into space right now, and you have *some* money, you can opt for an inexpensive little solid-fuelled rocket just 20 feet long and 10 inches in diameter. Question: when did you last meet a rich meteorologist? Well, exactly. Yet these humble weather watchers have been firing 'sounding rockets' into the sky since the 1940s, so that instruments in the nose cones can measure radiation levels and atmospheric densities and temperatures, essentially grabbing a quick snapshot of conditions in the high atmosphere before the rocket falls back to earth under a parachute.

(The name given to this type of rocket harks back to the days when sailors checked the depths of water under their ship's hulls by dropping weighted strings and then measuring how much they had to unreel before the string went slack, because the weight was touching the seabed. This was called 'sounding'.) The heights and trajectories achieved by these missions are modest and undramatic in comparison to a true space mission, and the fall back earthwards doesn't involve any great blaze of re-entry. Thousands of sounding rockets have had their brief moment of glory over the years, at costs that even the most impoverished scientific teams can handle. And the paperwork is cheaper too. Aviation authorities tend not to fuss about sounding rockets so long as they are fired from sensible locations well away from busy aviation routes.

Some private entrepreneurs are wondering what kinds of payload a simple-sounding rocket might carry that could turn a profit. In April 2007 the Connecticut-based company UP Aerospace made the first successful flight of its SpaceLoft XL rocket, carrying a 100-pound payload less than a foot in diameter and seven feet long. Hundreds of students from around the world developed 44 scientific experiments for the mission, compactly integrated into the rocket's slender shell. Even more compact were the 200 fee-paying passengers. Those who might have relished a space flight, only to find their ambitions cheated by death, at least had the satisfaction of letting their relatives give them a good send-off. Little canisters of cremated remains were loaded aboard the rocket and shot into space from the fledgling Spaceport America launch site in the desert of Sierra

County, New Mexico, 45 miles north of Las Cruces city. (More about Spaceport America later.) The rocket travelled 73 miles into the sky, just high enough to count as entering space. Then, after a four-minute suborbital arc, a parachute recovery system brought the 'cremains' down to earth for a more conventional interment. Among the cremated star passengers on this occasion were the pioneering NASA astronaut Gordon Cooper and the *Star Trek* actor James Doohan, better known as Scotty, the dauntless warp-drive engineer.

The only hitch in an otherwise flawless flight was that the lost loved ones were temporarily lost again. The recovery capsule came down amid the rugged San Andres Mountains in south-western New Mexico. The landing zone was well within the security remit of the White Sands Missile Range, which tends not to welcome casual visitors, let alone mountain-rescue teams from private space companies. UP Aerospace was, of course, granted permission to mount a search, but the weather took a turn for the worse, and to such an extent that within a few days the search team members were looking nervously over their shoulders for tornados sweeping across White Sands. Ah, well. Stuff happens. The point is that you too can set up an inexpensive launch business using sounding rockets, marketing your capacity both to the general public and to the scientific community. Thousands of researchers around the world are keen to make measurements at high altitudes, but without spending silly money for a full-on space launch or orbiting satellite. And it's not just a question of 'settling' for this junior-league version of a space flight. Sounding rockets are our only means of directly accessing certain layers of the atmosphere that are too high for planes or balloons to reach, and at the same time too *low* for orbital satellites to pass through without encountering atmospheric drag (air resistance).

what's in a name?

You might find that some scientists or PhD students are a bit sniffy about having their precious instruments loaded next to the remains of dead folk, so here's a marketing trick that you may want to think about – as pioneered by no less a rocket authority than the great Wernher von Braun (1912–77).

By 1957 the German-born visionary had built a little missile called *Redstone* for the US Army capable of sending a small satellite all the way into space, but the US government was embarrassed by him because he was an ex-Nazi (he designed the V-2 rocket). As a result, the first American space-rocket project, known as Vanguard, was awarded to an all-American Navy laboratory team, while von Braun was specifically ordered *not* to try to launch any of his *Redstone*s into space. In October 1957 Russia's *Sputnik* beat Vanguard into orbit and a few weeks later Vanguard, already a poor second in the Space Race between the US and the USSR, blew up on the launch pad. 'KAPUTNIK! flOPNIK!' said the newspapers. Belatedly, von Braun was asked if he could fix the mess, and quickly

– but without using a military rocket, because President Eisenhower had promised the world that America's first steps into space would be a peaceful civilian adventure. Von Braun had managed to stash a single *Redstone* in a hangar on reserve, waiting for his big day to come, but it had US ARMY painted all down the sides and the newspapers knew perfectly well what a *Redstone* was for. It was for dropping bombs on Russians. So what was he supposed to do?

Simple. He painted over ARMY and gave the rocket a new name. *Redstone* became *Juno* and it hurled *Explorer 1*, America's first satellite, into orbit in January 1958. Similarly, you can fire your commemorative stamps and coins and collectable knick-knacks and cremated loved-ones into the sky in one rocket 'uniquely' designed to carry payloads for the general public, and you can haul experiments aboard another tailored 'specifically' for serious science missions – and tarnation if they ain't just the same rocket painted in different colours! That's what history is for. You can learn from it. For complex political and funding reasons space agencies use this gimmick quite often, renaming rockets from flight to flight when, in truth, only small changes have been made to the actual hardware. You, too, can satisfy a range of different users or clients with one simple-sounding rocket.

a gentle climb

NASA knows all too well the difficulties of 'Max-Q'. That's the point in a rocket's flight when it reaches supersonic speeds while it's still in the low atmosphere, and the air in front, densely compacted by the shock wave created by the rocket itself, can't get out of the way fast enough. A massive amount of engine thrust is essentially wasted in just pushing past that barrier. Of course, once the rocket has gained more altitude and the air thins out the problem disappears, but ideally you don't want to burn up gigantic quantities of fuel simply fighting that resistance, so your challenge is to reach the thin upper atmosphere as painlessly as possible. The gentlest and most fuel-efficient way of getting your ship off the ground is to hoist it into the air underneath a gigantic helium balloon. You ascend slowly but surely to a giddy altitude of 70,000 feet or more, yet without once having ignited your rocket engines. The balloon does all the work, lifting your ship into the thin upper atmosphere nice and slowly. When you finally disconnect the cables between the ship and the balloon and fire your rocket engine you've already completed most of the hard work of the ascent. High-altitude balloons are relatively safe to operate, because helium is chemically inert and can't explode, even if you fire a flame-thrower into it.

As always, no system is perfect, including balloon launch. The canopies are extremely large and the thin fabric tears easily. A sudden gust of wind can cause havoc if the balloon is half-inflated but hasn't yet lifted entirely clear of the ground. Another problem is weight. How big do you want your spaceship to be? A six-seater? Five? Three? Actually the carrying capacity of a helium balloon is limited, unless it's going to be the size and volume of a sports stadium. Accordingly your ship will be relatively small and light-weight so as not to overstrain the balloon. That means you can forget the six-seater. In fact you had probably better not think about pas-sengers at all. If it's a space-liner business you're after, then balloon ascent may not be the best route.

Another problem is controllability. The balloon drifts as it climbs, and the whims of wind and weather will mean that you won't be able to calculate with absolute precision what point over the earth's surface the balloon will reach when you cut the cables and fire your ascent rockets. NASA and the other big space agencies fret constantly over 'launch windows', the carefully timed dates, hours, and sometimes even minutes when a rocket has to be launched in order for its payload to achieve a particular orbit or space rendezvous. If you want to launch a commercial satellite the inac-curacies of a balloon ascent to firing altitude may not be acceptable to your clients. On the other hand, not all space payloads have to reach a precisely specified orbit. Certain microsatellite science projects, especially those funded by small university and college teams, may be flexible about the exact region of orbital space you get them to, just so long as you get them there for a modest budget.

The other thing that you or your clients might want to do from a balloon hovering at the edge of space is to jump off it. This doesn't have to be as crazy as it sounds. After all, the logical extreme of a competitive parachutist's ambitions is to fall from as great an altitude as the laws of physics allow, and live to tell the tale. The current world record was set in 1960 by Captain Joe Kittinger of the US Air Force, who jumped from 20 miles high, briefly reaching a speed of 700 miles per hour as he fell, with almost no air resistance to slow him down – although there was enough atmospheric friction to warm the outside of his suit. He needed a cooling garment and several insulating layers to protect himself. Then, as the air thickened, he adopted the freefall parachutist's classic open-armed stance, using his body as a crude aerofoil, translating some of his downward rush into forward motion so that he could lose speed before at last opening his 'chute. It took Kittinger 13 minutes to reach the ground.

Modern helium balloons designed for air launch could also lift small gondolas suitable for parachute jumpers. We can certainly expect teams to try and beat Kittinger's record in the next few years. They'll also be able to test out new spacesuits suitable for the private market. 'Space diving' could become a popular TV sport; and wherever sport and media combine, you have a potential source of revenue. But these jumps don't have to be mere adventurism. There are serious purposes too. For instance, one of Kittinger's objectives was to test an early kind of spacesuit. Both NASA and the US Air Force have always wanted to know if it's possible to rescue astronauts from high altitudes; and today, one of the leading proponents of space diving is Jonathan Clark, a former NASA flight surgeon and military high-altitude parachutist, whose wife, Laurel, died in 2003 when the space shuttle *Columbia* disintegrated during re-entry.

fine-tuning

A more accurate air-launch system has been developed by the Orbital Sciences Corporation in Dulles, Virginia, one of the early success stories in the private space-flight revolution. *Pegasus*, a small rocket shaped like a cruise missile, drops from under the wing of a Lockheed L-1011 TriStar jet aircraft precision-piloted to 39,000 feet, then blasts into space for a fraction of the cost of ground-launched rockets. Pegasus rockets have flown more than 40 times in the last decade, boosting into orbit 78 satellites at a relatively cheap $14 million a shot – a quarter of the price of any booster available elsewhere. In a similarly fine-controllable technique developed by AirLaunch in Washington State, a small drogue parachute is released from the rear door of a military transporter plane and pulls a rocket clean out of the cargo hold. Once a safe separation distance is achieved, the rocket ignites and flies up into space.

Another company, Transformational Space (or t/Space) of Reston,

Virginia, has added another little trick to the air-launch business. Its prototype rocket or Crew Transfer Vehicle – known as the CXV and funded by a relatively modest $3 million research grant from NASA – drops from a carrier craft called *Proteus* (one of many elegant planes to have emerged from Burt Rutan's Scaled Composites company). On release, a cable attached to *Proteus* holds onto the CXV's nose cone for a fraction of a second, causing the little rocket to drop away stern-first. This ensures that the CXV is already pointing in the right direction – upwards, towards space – when it lights its engine. T/Space now want to look at larger rockets, lifted into the air by jumbo-sized carrier planes. Given that Charles Duelfer, the company's CEO, is a wily former intelligence analyst with a bulging governmental contacts book, t/Space has a fair chance of succeeding. It's even possible that a larger CXV could carry human payloads one day soon. But air-launch remains a new approach. You might prefer to put your money into a more traditional and well-understood launch technology, like a classic LOX/kerosene booster fired from the ground.

the classic route

PayPal founder Elon Musk's company SpaceX is developing a rocket called Falcon. In this case, 'developing' doesn't mean wandering around a Holiday Inn sales convention with armfuls of glossy sales brochures. Falcon is a fast-maturing machine today, close in power and size to an entry-level space agency-class launch vehicle. In March 2007 a Falcon reached space successfully, and although its upper stage failed to reach orbit as planned, there was little doubt in observers' minds that subsequent missions will succeed. The first payload was a 40-pound satellite from the Defense Advanced Projects Research Agency (DARPA), funded by the US Department of Defense. Although everyone was disappointed to lose that satellite, SpaceX has secured future launch orders from both governmental and commercial users. Falcon 1 is 70 feet tall and weighs 85,000 pounds fully fuelled. SpaceX offers to put payloads into low earth orbit for $7 million: a sizeable sum for mere mortals, but extremely cheap compared with any equivalent space-agency rocket. The much larger next-generation Falcon 9 is now under construction. It could lift space capsules and cargo modules to support missions to the International Space Station. At $35 million a throw, these flights would have to be paid for by NASA, but the point is that the space agency won't be in charge of *designing* the rocket.

the ultimate ride

SpaceX plans to deliver four astronauts into orbit aboard its Falcon 9, but who says that a passenger spacecraft has to carry more than one person – or even reach full orbit? Since the earliest days of aviation, aircraft manufacturers have catered to the private solo flier as well as the global passenger-carrying airlines. Boeing's jumbo jets and Airbus's leviathans are not the only winged vehicles capable of turning a profit. A vast range of tiny single-seater aircraft exists for sale to private pilots. How about a tiny single-seat spacecraft for sale to particularly dedicated space fans, or available for lease on a flight-by-flight basis? Here, the balloon-ascent mission profile might, after all, be right for your company, because the drift won't matter too much if all your client wants is the ride.

Indeed, the ultimate experience for any space traveller would be to fly in a spacecraft designed exclusively for one privileged private passenger: a person so devoted to the cause that he or she would be capable of funding a mission dedicated to their own fulfilment. Such a flight would be, in essence, almost a repeat performance of the first manned space flights undertaken by the USSR and America in 1961, when cosmonaut Yuri Gagarin and astronaut Alan Shepard became the first people in history to go beyond the earth's atmosphere and enter space.

To be more specific, it'll be Shepard's flight that the first solo private passengers will be emulating. Yuri Gagarin took off on 12 April 1961 at the tip of a vastly powerful rocket (by the standards of the day) called the R-7 or *Semyorka*. We've already talked about that beast. It was pretty much the same booster that had launched *Sputnik* in 1957 and some 50 years later it's pretty much the same booster that hauls Russian cosmonauts into space today. Basically the R-7 has enough punch to put a sizeable capsule all the way into orbit. But for the time being this gargantuan level of power isn't available to small, privately funded rockets. So let's talk about Alan Shepard's NASA flight and see where it differed from Gagarin's.

The truth is, Gagarin whizzed right around the earth in a flight that lasted 108 minutes, while Shepard's capsule splashed down into the Atlantic a mere 300 miles from his Florida launch site after a mission lasting less than 16 minutes. NASA's first man-carrying rocket, the *Redstone*, was tiny in comparison to the R-7, and couldn't get Shepard all the way into orbit. His mission, although greatly pleasing to the American public, was just a cannonball stunt, a ballistic arc. In other words, he flew exactly the kind of suborbital trajectory that most of the budding space-tourism industries are currently trying to exploit, because *sub*orbital is what you get when you use *small* rockets. Even so, it would be a special kind of thrill to re-enact Shepard's historic flight. The technology may have stretched NASA's greatest minds back in 1961, but today the rocket power and capsule heat-shielding systems are well understood and easily reproducible.

reinventing the 'vengeance weapon'

One project under development today is based on a notorious design dating back to the 1940s. It's the V-2 or 'Vengeance Weapon' built by Wernher von Braun for the Nazis during the Second World War. Standing 54 feet tall and steered by fireproof graphite vanes at the bottom that deflected the rocket exhaust, the V-2 was a stunningly effective piece of kit, although it wasn't much admired by the London residents whose houses it flattened or the slave labourers who were forced to assemble it. Its design influenced NASA's giant 1960s space projects, but of course America needed – and still needs – much larger rockets for its heavy space payloads. In contrast, a company called Canadian Arrow is building the *smallest* rocket that can feasibly carry a human and it's basically the V-2, albeit designed with modern lightweight materials and computer guidance systems, with an impressive engine at the business end delivering 57,000 pounds of thrust.

In 1946 the British Interplanetary Society (BIS), an educational lobbying group which included among its membership a young but already brilliantly far-sighted Arthur C. Clarke, came up with a detailed technical plan based on the use of a V-2 recently captured from the defeated German heartland. How about putting a specially trained RAF pilot into its hollow nose cone and launching him into the sky? The weight inside the rocket *not* taken up by an explosive warhead could easily be replaced by that of a man, plus the weight of the specially adapted detachable cone and parachute. Great Britain would then become the first nation on earth to send a man into space. With its unerring knack for failing to embrace the technological future, the British government rejected this relatively inexpensive idea, and this is exactly the suborbital flight profile flown by Alan Shepard in 1961, using a *Redstone* designed by von Braun and closely modelled on the V-2.

Take-off in a traditionally styled rocket is vertical, and fast. If you're the lucky astronaut, you will experience a tremendous surge of acceleration. At the top of the arc, some 60 miles into the sky, the capsule separates from the rocket and you will experience about five minutes of weightlessness. If there is a downside, it's that the capsule for this kind of flight has to be small and cramped, so there won't be much chance for you to unstrap your seat harness and float about to enjoy the sensation. Then, all too soon, your little capsule turns about-face so that its blunt heat shield is pointing in the direction of travel. The ghastly feeling of heaviness returns with a vengeance as the atmosphere slams once again into your craft and deceleration kicks in.

Two final nasty lurches await you. The first one comes at around 9,000 feet when the parachutes suddenly pop out, triggered by an automatic air-pressure sensor. You'll feel a jolt rather like the whiplash you get when some idiot drives into the back of your car, but the design of your couch should prevent your head from snapping off. The second and even nastier shock comes when your capsule hits the ground, the thud of its impact softened but by no means entirely cancelled out by those 'chutes. Russian capsules come down with a thud on dry land, but American capsules prior to the shuttle era always landed on water to soften the impact. Likewise, Canadian Arrow is opting for a splashdown to save the spines of its clients.

In 2006 the province of Nova Scotia signed an agreement with the US-Canadian company PlanetSpace allowing it access to 300 acres of government-owned territory overlooking the Atlantic, on which to construct a rocket-launching facility. Canadian Arrow needs this coastal site because the arc of its trajectory has to end over water; and in keeping with NASA's giant Florida complex, an eastern seaboard location gives some east-west momentum to the rocket as it climbs, helping it to pick up speed (as we've discussed – although the distinctly non-equatorial site at Nova Scotia reduces the benefits slightly), while ensuring that nasty mid-air disasters happen over water and not over people's houses.

Single-seater rides are all but inevitable in the coming decade or so, but in the shorter term the capsules will be fitted with up to three seats so that the manufacturers can sell enough tickets to make the system pay. The weight penalty imposed by multiple passengers will mean that the rockets can't climb very high, but so long as they reach that magic 60 miles into the sky, no one will complain.

winging it

The science-fiction dream *par excellence* would be a craft that simply whooshes into space, then comes home and lands again on a runway, just like any other plane. In the 1950s most space engineers imagined that a rocket plane would have huge 'delta' wings. Basically, the ship would look like an enormous flat triangle bisected by a needle-shaped fuselage. Today we know that when it comes to space flight wings are as much a hindrance as a benefit. They are great for lifting your ship off a runway and great for flying back to a landing, but in the vacuum of space they are parasitical 'dead weight', because they have no job to do; and during the first and most intense phase of atmospheric re-entry they can be a nuisance. If the wing is thin, which it wants to be if it's going to be any good in the atmosphere, then its leading edge will get very hot during re-entry, because it's difficult to diffuse the heat away from that small and sharp front surface.

You could try to design the rocket plane's wings to be so discreet that Mother Nature won't even notice them. In 1959 NASA built an experimental rocket plane called the X-15. It was shaped like a black dart with an enormous rocket nozzle at one end, a thick tailplane and very short wings, added almost as an afterthought to give the pilot some hope of controlling his touchdowns. It took exceptional piloting skills to handle a craft whose wings were as small as they could be without actually becoming ineffectual. The X-15 could fly at more than six times the speed of sound at altitudes of up to 350,000 feet, where the air is so thin there essentially isn't any. Six times the speed of sound, by the way, is about a mile per second. This was a *very* fast machine and just like the *Pegasus* spacecraft it was carried aloft by a jet aircraft (a converted B-52 bomber) and then released at around 50,000 feet to begin its rocket-powered climb to space.

Harrison Storms, the X-15's charismatic designer, pleaded with NASA to let him 'beef it up a little bit', so that it could go all the way into orbit. Getting up there wasn't the problem. It was coming down again that clipped the X-15's ambitions. Orbital velocity is 23 times faster than the speed of sound or 17,500 mph. Once you've used up all of your rocket fuel to achieve this giddy rush, how are you supposed to slow down again and descend back to earth? The X-15's outer skin was built from an exotic alloy called Inconel-X, a material capable of withstanding temperatures of up to 1,000° Fahrenheit (537° Celsius). Unfortunately, the plunge back into the atmosphere at only slightly less than 17,500 mph would have heated the X-15 to more than three times that temperature, and no one in the early 1960s knew how to protect the little ship and its pilot from such extremes.

capsules and pods

In the 1960s rocket boffins decided, albeit reluctantly, that conventional wings were just not suited to the challenges of re-entry from full orbital velocity. As we all know, NASA and its Soviet rivals both adopted the 'capsule' approach to space flight for their pioneering astronaut and cosmonaut missions. Blunt, round-shaped pods or soft-edged cones were swathed in thick layers of shielding, which was allowed to burn away (or 'ablate') during re-entry, dissipating the heat while protecting the crewmen inside the pods from being vaporised. It worked, after a fashion, although the exterior skin of a capsule ended up scorched and scarred almost beyond recognition once it came back to earth.

Obviously the skin of a re-entry spacecraft is heavily shielded against the heat, but much more important is the shock wave or 'bow shock' of compressed air that the craft makes as it hurtles through the atmosphere. The streamlined shape of a typical aircraft is designed to reduce atmospheric drag by minimising bow shocks at normal flying speeds. That's why most jet planes and airliners look so sleek. But at re-entry speeds of several thousands of miles per hour you want that bow shock as thick as possible, partly to help slow you down, but just as importantly to act as an insulating layer between the skin of your ship and the furious friction of the surrounding atmosphere. Ordinary wings are a nuisance, because their bow shocks are so narrow they don't create a decent insulation layer, so they'll just burn away.

If you do want those wings, one option is to fatten them up and round off their leading edges to create a good bow shock. This is the method adopted by NASA's space shuttle. If you look at its wings you will see that they are dramatically thicker and bulkier than those of a normal plane. Now comes the drawback. Thicker wings need more materials for their construction and this means that your spacecraft's take-off weight rises, so you need more fuel and more powerful engines to get it off the ground. We've been here before, haven't we? The vicious circle of increasing weight. The shuttle is a heavy beast requiring a vast amount of rocket fuel to get it off the ground, largely because of those big fat wings.

The shuttles were designed in the late 1970s at a time when it was almost inconceivable to build a spacecraft out of anything other than aluminium or titanium, with maybe some ceramic heat shielding glued to the outside. The colour scheme of the shuttle is a handy guide to its insulating materials. The brittle white ceramic tiles on the fuselage and the tops of the wings stay fairly cool during re-entry, because the air flows past them obliquely. By contrast, the shuttle's nose and underbelly hit the atmosphere pretty much head-on. The black tiles on these surfaces are crucial, but the grey strips on the leading edges of the wings absorb the most severe punishment of all. They're made from reinforced carbon-carbon (RCC) panels capable of surviving temperatures of 3,000° Fahrenheit (1,648° Celsius). The double use of the word 'carbon' denotes the different forms of that element involved in the mix: graphite for heat absorption combined with a webbing of thin carbon fibres for strength.

RCC panels are much stronger than ceramic tiles, but the drawback is their brittleness. If something hits them hard enough, they will crack. They can survive a fantastic amount of re-entry heat, but have very little flexibility or 'give'. In the *Columbia* disaster of 2003 a suitcase-sized piece of thermal insulation foam peeled off the shuttle's huge external fuel tank shortly after launch and hit the left wing's leading edge, cracking one of the RCC panels and making a small but ultimately catastrophic hole. During re-entry three weeks later hot gases rushed into that hole, destroying internal controls and melting the underlying metal airframe. The shuttle disintegrated and the crew died.

the shuttlecock

Spaceships returning from orbit have little choice but to meet the re-entry challenge more or less head-on, because they're barrelling forwards at full whack, but a *sub*orbital craft has a more gentle option: it can lose most of its speed on the way up. This might sound crazy after all the fuel and fury you've devoted to the rocket-powered climb, but think of it this way. You want the rocket engine to hurl you upwards, but you don't need very much forward momentum, because it's not your intention to stay up in space for more than a few minutes. So you point your craft as vertically as you can during the rocket surge, and then, when the firing phase is over, you let gravity reassert itself so that your upward momentum is lost again.

It's just like throwing a tennis ball in the air. At the top of its arc it hovers for the briefest moment, caught precisely between the force of gravity pulling it downwards and the last dregs of upward momentum imparted by your throwing arm. Then gravity wins completely and the ball falls back down to earth. The point is, you threw that ball upwards as hard as you could, didn't you? Yet it slows almost to a standstill at the highest point of its trajectory. According to the Mojave-based company Scaled Composites and Burt Rutan, its chief designer, this is the kind of flight profile you need for a suborbital craft, so that its fiery descent back into the atmosphere begins as slowly and as gently as possible. Rutan's X Prize-winning craft SpaceShipOne (SS1) was a small-scale prototype for the larger ships that Virgin Galactic will soon bring into service. Let's see how the system works, using SS1 as an example.

SS1 has sometimes been compared to the old X-15 by virtue of the fact that both machines were spaceplanes flying suborbital trajectories. Actually SS1 had relatively little in common with its ancestor. X-15 was built for extremely high speed, while SS1's goal was pure altitude. It could shoot almost straight up and fall back from space almost straight down at relatively low speed, whereas the X-15, travelling much faster, took more oblique trajectories that sent the craft hundreds of miles 'down range' to land. SS1 was made of graphite composite materials and weighed only 7,000 pounds, while the X-15 was built from titanium metal alloys and weighed ten times as much. One ship slammed through the atmosphere like a pointy shaped brick hurled from a catapult, the other buffeted gently against it like a feather dropped from a tall building.

SS1's first trick was to use a carrier plane (known as the *White Knight*) to carry it to its release altitude. *White Knight*'s wings were very long, like a glider's, to maximise the plane's interaction with the air, while its jet engines were lightweight and relatively low-powered, because the wings did most of the work. It may surprise you that the velocity for this phase of the journey seldom exceeded about 380 mph. Raw speed wasn't the issue. The atmospheric ascent was all about getting that potential blockage of air resistance safely out of the way without fuss, just like the other air-launch techniques we looked at earlier. At 48,000 feet *White Knight* levelled out and used its wings to get the best grip on the last thin traces of air. In this border territory between the atmosphere and the vacuum of space the carrier plane couldn't ascend much higher, even with its unusually long wings. Now came the moment for SS1 to be dropped from under the *White Night*'s belly, fire its rocket engine and pick up fantastic speed without having to worry about drag or Max-Q.

SS1's rocket engine burned for 80 seconds, accelerating the craft at three times the force of gravity, achieving a speed of Mach 3.2 (2,100 mph) by the time the engine burned out at 160,000 feet. The ship then coasted, unpowered, on momentum alone, to the top of its arc, now *losing* speed because the engine had been switched off. At 200,000 feet the pilot started to experience weightlessness, which lasted for about three minutes as the spacecraft reached its maximum altitude of 340,000 feet (62 miles). At this point the ship came to a virtual standstill for a moment, its upward momentum precisely cancelled out by the remorseless tug of gravity pulling in the opposite direction. Now, at the highest yet slowest part of its trajectory, SS1 used little low-powered 'cold-gas' thrusters to nudge its nose this way and that, point the windows in the right direction for a good view, and align the craft properly for re-entry. Then gravity reasserted itself and the ship fell back to 200,000 feet, when it began to feel the drag of the atmosphere once again. SS1's next trick was to change shape.

'Variable-geometry' aircraft have existed since the Second World War. Navy fighter planes such as the Wildcat and Corsair folded their wings to save space on the cramped decks of aircraft carriers, while Germany's Messerschmitt P.1101 experimental jet fighter could alter the shape of its wings mid-flight. In the early 1960s the F-111 jet fighter-bomber employed 'swing wings'. For take-off and landing at low speeds, the wings stretched out to obtain maximum lift, while at supersonic speeds they folded tight against the plane's fuselage to reduce air resistance. Modern F-14 Tomcats, RAF Tornado fighters and Russian Tupolev bomber jets also have swing wings.

SpaceShipOne exploited this old trick in a new way. With the help of hydraulic pistons the craft folded its wings perpendicularly to the fuselage instead of parallel, then gently fell back towards the earth like a shuttlecock in a game of badminton. The steep, almost vertical angle of attack dramatically increased the drag and slowed the descent. The ship came down belly-first rather than nose-first, massively increasing its bow shock. This so-called 'feathering' technique enabled the craft to align itself belly-down against the atmosphere automatically, with very little need for the pilot to wrestle with the controls. SS1's maximum descent speed was a startlingly modest 1,900 mph, so the frictional heat was much less than, say, for an X-15 barrelling back in at several times the speed of sound. Finally, at 80,000 feet, the SS1 pilot reset the wings to their normal horizontal position. The ship became a glider and coasted home to a runway landing.

And yes, as always, there were difficult moments in the SS1 story. There were some alarming control problems at transitionary speeds, altitudes and air densities: at those points, basically, where SS1 wasn't quite sure if it was a rocket or a plane. In some of the test flights the ship rolled sickeningly from side to side as the wings clutched ineffectually at the last traces of air on the way up. The cold-gas thrusters wouldn't have made much difference during the powered rocket ascent, because their small contributions would have been swamped by the greater forces of the main engine. While that did its job perfectly well, pushing SS1 upwards like a bat out of hell, the system for stabilising roll motions was not especially good. If all this sounds rather academic-aerodynamic, think of an arrow shot from a bow. In the air the flight feathers at the back make the arrow's flight much more stable. In a near-vacuum these flight feathers won't get much of a grip and any slight twist imparted by the bow string's thrust eventually causes the arrow to wobble.

Given the many different roles that a spaceplane's hardware has to play in a mission – from plane to rocket to re-entry vehicle and then back to plane again – there will always be some ambiguous phases in a spaceplane's journey when nothing is quite perfect, and some essential degree of risk enters the equation. Finally, it has to be pointed out that SS1 wouldn't have been strong enough to survive a return from orbital velocity. That problem is orders of magnitude tougher.

all-in-one

The logical next step is to reduce the carrier plane and spaceplane combination to just one ship. An 'integrated' spaceplane takes off exactly like the executive jet which it so resembles in appearance. Twin jet engines at the rear of the craft provide the power, but they don't have to be heavy fuel-guzzling monsters, because the long slender wings do most of the work, lifting the vehicle off the ground and allowing it to soar to high altitudes, essentially as a powered glider. Your power-to-weight ratio is a good one, and your fuel efficiency is also maximised. From the passenger's point of view this slow and conventional take-off delivers the most comfortable ride, no more disconcerting than the jolt they would experience in a conventional airliner. Even so, there must come a point, at an altitude of around 50,000 feet, when the air thins out, the jet engines are powered down and the rocket nozzle at the rear of the fuselage kicks in.

When this engine fires up the acceleration will be sudden and dramatic. What's more, the ship's angle of attack will increase. In other words it will climb upwards at a very steep angle. Now your passengers will feel the force of the acceleration pressing them deep into their seats, and this time the sensation will be markedly more intense than anything they will have encountered previously in a conventional plane. Your seat designers, therefore, are faced with an interesting challenge: how to maximise the comfort of passengers for a gentle horizontal take-off, but also for a stomach-lurching acceleration almost straight upwards into the sky; and then, of course, how to perform the same trick in reverse, allowing the passengers to absorb the leaden feeling of deceleration as your craft re-enters the atmosphere after its brief sojourn in space.

The engineers at EADS Astrium, a major European manufacturing company, have designed a seat that swivels moment by moment with each phase of the journey, so that the passengers' backs are kept always in the most comfortable position to absorb the stresses of acceleration and deceleration. Instead of facing forwards towards the nose of the craft, the seats are slung crossways like hammocks. As the plane pitches its nose dramatically up and down, each seat's angle is constantly adjusted to minimise the discomfort. The unusual sideways mounting does mean that passengers will not have 'window seats' in the traditional sense of that phrase. The windows on the left and right flanks of the fuselage will be above their heads or below their feet.

EADS Astrium is more than capable of delivering this suborbital craft. It builds the powerful Ariane rockets, workhorses for the European Space Agency. Even so, the problem remains that no spaceplane with long slender wings can make it back from full orbital speed. With that higher and greater prize in mind, some designers think it's better to abandon wings (almost) entirely and design a blunt, softly rounded capsule that can come home like a plane.

lifting bodies

The Mercury, Gemini and Apollo crew capsules of the 1960s were perfect for re-entry, being round and blunt to shrug off the heat, but they couldn't fly to controlled landings on a runway. In fact they couldn't even survive hitting the ground. So they were recovered after splashing down in the Atlantic Ocean. Many NASA personnel thought that this was a rather undignified way for astronauts to come home. Dale Reed wanted to land them on a runway, but didn't want the problem of wings with sharp leading edges and complicated heat shielding, because, as we've seen, they get very hot on re-entry. What was needed was a capsule that could fly *like* a plane, but not actually *be* a plane. This was what Reed called the 'lifting body' concept: a fat round-edged triangle, shaped like a soft wedge of cheese, that has no sharp leading edges yet still creates aerodynamic lift. The senior managers in NASA didn't believe a machine without wings could fly, so Reed built a wooden prototype for $10,000, essentially small change that he didn't have to get approval from head office to spend. Then he and his colleagues towed the vehicle behind a car, a souped-up Pontiac. The first pilot, Milt Thompson, controlled it beautifully.

By the summer of 1963 Reed's wooden lifting body M2-F1 (nick-named '*The Flying Bathtub*') was reaching ever higher in the air, towed behind an aircraft and then released. It performed amazingly well, and these initial experiments convinced NASA and the Air Force to fund much more sophisticated metal lifting bodies capable of high-speed high-altitude flight. They were carried 70,000 feet into the sky, tucked beneath the wing of a B-52 bomber. And then they were dropped. Given that these ships were unpowered, their pilots had only one chance to touch down safely. Lifting body research was vital for the development of NASA's Space Shuttle, although, as we've seen, that ship is still more of a conventional glider than a lifting body. It does still have huge wings, even if they are softly rounded at their leading edges according to lifting-body principles. Apart from the political wrangling throughout its design, the shuttle was simply too heavy to land without those wings.

For the time being at least – and for reasons that we'll investigate in more detail later – NASA has abandoned runway landings for the future. The spaceships scheduled to fly after the shuttle is retired in 2010 will be surprisingly similar to the old Apollo capsules: nothing like the shuttle, and not quite lifting bodies either. But a number of private space companies are not yet willing to give up on the lifting-body concept, and you might find something of value in the reams of test-flight data available from the 1960s.

single stage to orbit

Lifting bodies would still need a separate booster to get them up into space. Another important concept is known as Single Stage to Orbit (SSTO), a fully reusable craft that doesn't shed boosters and fuel tanks as it rises. Back in 1996 NASA announced a competition to design shuttle: the Next Generation. In July 1996 the Lockheed Martin company won with its distinctive wedge-shaped concept, known as VentureStar, based substantially on lifting-body ideas. The prize on offer? $1 billion of development funding for X-33, a smaller unmanned prototype. Rival corporations were surprised, because out of the three rival bids the Lockheed concept was the least mature. But Dan Goldin, NASA's chief at that time, wanted something truly radical to pep up his agency's ailing technology base. 'We haven't developed a genuinely new rocket technology in 20 years, and that's a disgrace,' he said, with characteristic forcefulness.

Anyway, with their winnings in the bank Lockheed spent three years in a bold attempt to reinvent space transportation from top to bottom, starting with the X-33 prototype, a single fully integrated piece of hardware shaped like a round-edged triangle. Weighing in at 130 tons and standing 70 feet tall this hefty vehicle was almost exactly the same configuration as the planned VentureStar, but half the size. It was supposed to be the world's first genuinely reusable spaceship, with no external boosters or throwaway fuel tanks. It would launch vertically like a conventional rocket, then land horizontally, exploiting its lifting-body shape.

If you remember, we discussed earlier how a rocket needs differently shaped engines for different phases of an ascent. Dan Urie, the X-33's designer, came as close as a whisker to inventing a single engine that can perform all of the necessary roles. The 'linear aerospike' engine uses the earth's atmosphere to shape the exhaust plume. At low altitude, where the atmosphere is densest and presses in from all sides against the exhaust plume, the aerospike fires straight down, thus providing the most efficient thrust. But at high altitudes, where the atmosphere is thin, the exhaust plume tends to splay out at the sides and lose its efficiency, as we've seen. To compensate, the plume can be reshaped inwards using a curved metal reaction plate. It's as if an invisible nozzle made from the atmosphere on the outside and the reaction plate on the inside restrains the hot exhaust gases in a 'nozzle' whose outer walls (the air) are invisible. The burn is shaped for maximum efficiency at every moment of the climb.

Finally, the X-33's overall weight was supposed to be dramatically reduced by using carbon-composite materials instead of metal for much of its construction. Unfortunately the design of the experimental ultra-lightweight fuel tanks was too ambitious. They could not hold their liquid-hydrogen fuel without springing leaks. Lockheed fitted simpler aluminium tanks instead, but the additional weight made X-33 too heavy to fly. The entire project was cancelled in 2001. Your hot tip: take another look at the one component that did work flawlessly during ground tests – that aerospike engine.

in true buck rogers style

While not exactly gloating over the X-33's failures, a rival company, McDonnell Douglas, came up with the Delta Clipper, a vehicle that lifts off *and* lands vertically in true Buck Rogers style. The quarter-scale prototype was 30 feet tall and 12 feet across at the base, and shaped like a tall square-ended pyramid, with four rocket engines in its stern and four landing struts protruding from its flanks. It rose impressively on a column of fire and descended to a pinpoint landing on that same column. NASA turned its nose up at this idea. It wasn't happy about relying on rocket power to bring the ship home safely. Undaunted by NASA's rejection, McDonnell sought funds from the US Strategic Defense Initiative Organisation (otherwise known as 'Star Wars') and the Delta Clipper prototype began flight trials. Unfortunately in 1996 it confirmed the worst fears of its critics by catching fire immediately after touchdown when an engine sprang a fuel leak. It toppled over and was engulfed by flames. Yet this is the design chosen by Amazon founder Jeff Bezos and his Blue Origins company.

The trick with vertical take-off and landing (VTOL) ships is to make sure that your design features an even number of rocket engines at the base, each engine powered by its own subset of pumps and pipework. Then, if any single engine shows signs of failure, here is what you do: you shut it down. And just as immediately you shut down another one: specifically, whichever engine happens to be diametrically opposite the dud one on the base of your ship. This might sound crazy, but it is more important to maintain the *symmetry* of rocket thrust than the raw power of it, otherwise your ship will topple over and lose control.

A five-engine cluster can work for you, with four nozzles on the outside and one in the middle. If you lose that central engine, the outer four can continue to fire uninterrupted. In 1970 the giant

Saturn V rocket used for the Apollo 13 mission experienced this problem when the middle one of its five thunderous engines conked out shortly after launch. Naturally, the Saturn's designers had planned for the worst-case scenario. If an outer engine had gone south, then its diametrically opposed twin would have been shut down and Apollo 13 would have limped into the sky on a still-symmetrical cluster of three engines in a neat row. From that scenario, at least the astronauts could have made an emergency splashdown.

Actually, as you might remember, Apollo 13 made it all the way into space and its real problems (an oxygen tank explosion in the rear service module) only began when the capsule and its attached lander were well on the way to the moon. So how did that Saturn get all the way into space on only four out of five engines? Here's the secret, both for yesterday's Saturns and the VTOL ships of tomorrow. Your final engineering tweak is to design your engines to fire at 100 per cent thrust for most of their working lives, yet each should be sufficiently rugged to run at, say, 120 per cent power *occasionally*. This will enable you to boost their output in an emergency to make up for the failed engines and their shut-down twins. If your engine failure is relatively minor and non-explosive, you may well be able to complete your mission safely.

No backup system is flawless and an engine cluster will occasionally have a 'no-win' moment. The Soviet Union's gigantic and ultimately doomed lunar-landing booster, the N-1, was only supposed to go up, not come down again, yet it serves as a pointed lesson when it comes to VTOL design. It employed 30 engines mounted in a flat round base: 24 in the outermost ring and six on the inside. An electronic system known as KORD was supposed to maintain that all-important thrust symmetry if any engines sputtered out. In February 1969 the N-1 flew an unmanned test flight as the USSR tried one last time to try and catch up with America's Apollo project. After a thunderous launch the N-1 was climbing nicely when two engines on the outer ring suddenly failed. KORD compensated by switching off another couple, and for a while the ascent continued smoothly. Then a fifth engine's pipework caught fire and KORD was faced with a dilemma. How to keep everything balanced with yet another engine on the blink and too many failures occurring, now, on one side of the rocket? With perfect mathematical logic, KORD decided the only option was to shut down all the remaining engines . . . Oh, well. At least no one could say it hadn't kept everything symmetrical.

planning for problems

This business with the VTOL engines might help you to think about your 'abort scenarios'. Basically, what do you do when things go horribly wrong, but you want your crew to survive? Now, the thing about things going horribly wrong is that they can go horribly wrong at any given moment. You have to plan for all possibilities. Here's a brief list of things that can go horribly wrong. Barring a game-over explosion from the get-go, what should you do when your ship

(1) sits on the launch pad fully fuelled, with your crew aboard, and the 'ignition' signal has been given but nothing happens?

(2) catches fire on the launch pad and threatens, possibly, to explode?

(3) launches successfully, but has a serious engine failure mid-ascent?

(4) gets high into the atmosphere, but doesn't quite make it into space because of an upper-stage engine failure?

(5) reaches space successfully, then suffers a life-support or power failure?

(6) loses control during re-entry or has a parachute failure or other landing problem?

For much of the Space Age both Russian and American manned rockets were tipped with needle-shaped 'launch escape towers': simple, solid-fuelled rockets that could lift crew capsules away from their boosters if anything went wrong during launch. September 1983 saw the first ever use of an escape tower in deadly earnest, when cosmonauts Gennady Strekalov and Vladimir Titov were pulled clear of a malfunctioning Soyuz booster. The automatic sensors failed to notice anything amiss, but a canny ground controller (who knew a launch pad explosion when he saw one) activated the escape tower's rockets by remote control. The cosmonauts had a spine-jarring jolt as the emergency solid rockets dragged them forcibly into the sky. Less than a minute later, the booster and its launch tower were engulfed in flames. By then the Soyuz capsule and its crew were already drifting back down to earth under their parachute, several miles clear of the disaster. Afterwards the cosmonauts, shocked and bruised but very much alive, said they had experienced 'a second birthday'.

NASA's Mercury and Apollo capsules employed a similar system, while the two-man Gemini capsule was equipped with large, gull-wing-style doors above each pilot, to enable spacewalking. Gemini's designers realised they could save the weight of an escape tower by lining the edges of each hatch with explosive bolts, enabling the crew to shoot out of the capsule on ejection seats. In the event, all the boosters worked and none of these escape systems had to be fired. Maybe that's why NASA lost track of reality in later years, becoming complacent about safety. The space shuttle has no escape system and there is no workable abort scenario for rescuing its crew of seven from the main rocket stack in the event of failure. That's not the kind of mistake that NASA can afford to repeat in its future spacecraft designs – and neither can you, unless you want a *Challenger*-style explosion to haunt your dreams forever. Always plan for trouble and assume the worst, from one split second of a mission to the next.

helicopter into space

So the VTOL concept is difficult to pull off safely, yet it tugs at the heartstrings of any space fan raised on the gaudy artwork covers of 1950s science-fiction magazines. Even the brilliant engineers of NASA's Apollo project were seduced for a while by the charm of the 'rocket ship', the tall cathedral spire of its needle-nosed fuselage propped upright by the slender buttresses of its tail-fin legs. But this beautiful cliché is a dangerous chimera. Apollo was going to be such a craft, until wiser heads spoiled all the fun by pointing out the difficulties of backing a tall rocket down to a safe landing on the moon. It would have been like trying to park a Navy corvette on its stern without having it topple over. The centre of gravity of such a tall rocket ship would be too high above the lunar surface, too prone to upsets. Notice how the eventual design for Apollo's bug-like Lunar Module had a centre of gravity tucked neatly inside the span of its four widely spread legs. It's difficult for it to fall sideways, like a drunk losing his balance. In fact it quite often landed on sloping terrain without so much as batting an eyelid.

Nevertheless, some folk remain wedded to the idea of tall space-ships flying VTOL missions. Bevin McKinney and Gary Hudson's machine, the Roton, is perhaps the strangest spacecraft ever designed. The main structure – with cargo bay, fuel tanks and crew cabin – is a conventional rocket approximately 60 feet tall, but the upper section is essentially a helicopter, complete with spinning blades. This eerie hybrid seems like some fantastical machine from a Dan Dare or Buck Rogers cartoon strip. Roton is a single-stage-to-orbit (SSTO) space launcher. There are no drop-away boosters or external fuel tanks and the entire ship is reusable. As well as using rocket engines in its base, the Roton is powered partly by small rockets attached to the tips of the four huge rotor blades windmilling from its domed peak. Kerosene fuel and liquid oxygen is pumped along the inside of the blades so that the rapid spinning motion creates a fuel-injection pump, forcing kerosene into the rotor-tip engines at high pressure. As the rockets fire up, the rotors whirl faster than the speed of sound.

At launch, the tip rockets are angled parallel to the ground, spinning the blades until they lift the Roton into the sky. In contrast with the mighty roar of a shuttle or a moon booster, the Roton's main engines operate at only a small fraction of their maximum power at ignition. The ship climbs out of the dense atmosphere using very little of its precious fuel. The helicopter blades do most of the work, using just enough thrust from their tip rockets to keep them spinning. As Roton climbs above 35,000 feet, the earth's atmosphere begins to fade away and the helicopter blades can no longer provide lift. The blades are gradually tilted so that the rockets in their tips can fire almost directly downwards, now at maximum power and nearly in parallel with the engines in the base. However there is still a slight sideways bias to the tip-engines' thrust. The blades have to keep spinning, even in a vacuum, to force fuel into the rocket chambers.

Once Roton has reached orbit, the blades are folded along the ship's sides. Cargo bay doors open, and the ship delivers its payload into space. Then it turns its blunt heat shield 'into the wind' and fires small braking rockets to slow it down. The Roton's fiery re-entry begins in familiar fashion, but once the ship has plunged into the denser atmosphere, the rotor blades flip out and start to spin again. This time it isn't the rockets in the tip causing them to rotate, but the air rushing past. The blades now act like a parachute, slowing the Roton's descent yet further. The pilots switch to manual control for the last 500 feet of descent, adjusting the angle of the rotor blades to land their ship very precisely, just like a helicopter.

Because of the noise of take-off, plus the risk of explosion, most VTOL rockets can only be fired from remote locations. But when the Roton takes off, very little rocket blast hits the ground. This eliminates the need for a launch tower and flame trench, so the spaceship can operate from a flat tarmac apron. After a mission, it doesn't need a runway for landing, so this is another huge cost saving in ground facilities. In an emergency, the Roton could come down safely in any farmer's field. In fact, safety is a key design feature. If anything goes wrong with the rockets during launch, they can be shut off and the Roton will make an emergency landing with its helicopter blades in 'parachute' mode.

At least, it *would* do all of these things if its designers, McKinney and Hudson, hadn't fallen prey to that most intractable failure of all: financial meltdown. The Roton prototype is on open-air display at a road intersection near Mojave Airport, a sad relic from a broken dream. Maybe you can rejuvenate some of the intriguing ideas behind its design?

the ultimate engine

Jet engines crop up time and again in our story, propelling the various carrier aircraft for (for instance) the Pegasus, the X-15 and the SS1. The EADS Astrium Space Plane uses jets before switching to a rocket engine for the final boost. But what if you could save weight and complexity by having an engine that works both in the air and in the vacuum of space?

In one sense, jets and rockets are similar beasts. You pump fuel and oxygen into a combustion chamber, put them under pressure, then ignite them. In a rocket engine, the pressure is applied by turbopumps within the fuel system, while a jet uses compressor fan blades inside the engine itself to force air into the system at pressure. A rocket carries its 'air' (the oxidiser) in an internal tank, while a jet exploits the oxygen freely available from the surrounding atmosphere. Jets, of course, run out of atmosphere at around 50,000 feet, and that's when rocket engines for an air-launched spacecraft usually have to kick in. However, an extremely fast jet (known as a scramjet) can continue to exploit the thin upper atmosphere simply by virtue of scooping it up so *fast* that the tiny traces of oxygen available accumulate rapidly enough inside the engine to keep it firing.

Scramjet technology has already been tested by NASA. In November 2004 an unmanned X-43 craft accelerated to nine times the speed of sound. Three years later, Australia's Defence Science and Technology Organisation (DSTO) flew its equivalent machine to an altitude of 330 miles and clocked up a speed of Mach 10. One day soon, scramjets will be able to switch from externally supplied air to an internal oxidiser without skipping a beat. A normal turbo-fan jet couldn't do this, because its fan blades kind of *get in the way* when the rocket combustion kicks in. The scramjet, by contrast, uses the sheer forward speed of the vehicle to force air into the engine, so there are no fan blades inside the firing chamber. The snag, iron-ically enough, is that you have to get your scramjet moving extremely fast straight away, before its engine can work properly, so a small conventional rocket stage may be needed to get the thing moving once the carrier aircraft has released it. The other consider-ation, of course, is that such colossal speeds are hair-raisingly dangerous, so this technology is probably best suited to the dark machinations of the military.

four 'no bucks,
no buck rogers'

One of the first things your investors and bankers will want to know
is this: how will your beautiful but very expensive spacecraft pay for
itself? Building spaceships is completely different from manufacturing aeroplanes or other high-performance vehicles commercially. In
the usual way of things, experimental prototypes – jet fighters for the
military sector, say, or airliners for the civil market – are shaken
down in a long series of test flights until a proven model can be
mass-produced for sale to the client. Unexpected challenges
inevitably crop up during the design phase, and the extra costs

incurred are absorbed into the unit sales prices of production models. That's called *amortising*. To give a somewhat over-simplifed example, if you create an incredible new bicycle, and it costs you $10,000 to perfect the design and tool up for the components, then you just amortise a discreet $10 into the sales price of the first 1,000 bikes to go on the market. The problem with most spacecraft is that there never really *is* a mass-production run, so it becomes difficult to recover a sensible proportion of your development costs from the sales price of each craft.

That shortish production run gives you another headache. Your first fleet of ships – numbering no more than half a dozen or so – is the finished product, and its missions into space have to be considered essentially as glorified test flights. Because, let's face it, who in the history of rocketry has ever flown *anything* into space and back so many times that its every last quirk and snag has been identified and ironed out? You won't find it easy to perfect the design of your machine until you've flown it dozens, perhaps even hundreds of times, until your engineers have had a proper chance to wrangle every last bug out of it. The average Boeing 747 airliner has lumbered up into the air so often, it takes something truly unpredictable – like a terrorist attack or a ground controller's error – to put one of those machines into a tailspin. Boeing's design has been honed and proven during countless flight histories, so that for each plane's purchasers and users (and just as importantly, its insurers) it is considered a reliable technology. It also helps that more than a thousand jumbos have been built since the first one rolled out in

1969, and not just four or five of them. The latest ones are incremental improvements on versions that worked before, rather than dangerous revolutionary experiments. There's not much about those planes that their engineers don't know. By contrast, there are plenty of surprises – and few of them particularly nice – that NASA's Space Shuttle seems to come up with every time it flies. That's because it's a rare and relatively seldom-flown experimental craft and not a proven *product*. Its total flight history amounts to fewer than 130 trips.

As far as the spaceship business is concerned, Russia's Soyuz capsules and boosters are the closest that anyone has come to a mass-production run. The giant ex-Soviet space factories just keep churning out those things, year after year, so these days the costs of each ship are fairly well stabilised by economies of scale. Hundreds of flights, manned and unmanned, almost all of which have been successful, have proven the booster system, while minor problems or averted disasters over three decades and more have helped Russian engineers refine the crew capsule to the point where most space experts think of it as trustworthy at least, if not exactly the last word in comfort. The downside to this success story is that the major components of the Soyuz hardware only last for one mission before being thrown away; and that means it's still an expensive ship in terms of getting passengers into space at a price you can afford. Each passenger seat on a Soyuz, remember, costs at least $20 million, because the passenger is basically helping to underwrite the construction of the entire ship, albeit from a set of well-established factory blueprints. But look on the bright side. At least when you order up a Soyuz no one is going to come to you eighteen months down the line and whinge that unfortunately the designers were a little too ambitious and the ship can't actually be built, so all your investment to date has gone down the pan, and no one will risk giving you any more money. At that point, game over. Or, as the old hands at NASA used to say, 'No bucks, no Buck Rogers.'

Ideally you'll be looking for a fleet of *reusable* spacecraft that can be flown reliably a hundred or so times before you scrap them, so that eventually you get your money back by accumulating passenger sales. This has never been achieved by any spacecraft company yet. Or else you might want to build dozens of vehicles that are only designed to last for one flight, while keeping the assembly price of each one as low as possible. Again this is not an easy call, unless you think in terms of the Soyuz and its matching booster rocket, co-financed by clients in the $20 million-plus range. All spacecraft will be unusual and expensive items for at least the next decade, until the technology matures. This is why suborbital space trips are being marketed at a still quite substantial $100,000 to $200,000 asking price, while anything in the orbital range will continue to cost tens of millions of dollars. Don't be downhearted. These sums are still a great deal less than the billions traditionally spent by the big space agencies. You still need to be rich to join in the space game, but just look around you. Quite a lot of folk *are* rich enough to take part. Aside from the fact that they are wealthy, they are people just like you.

adding value

If the technical development of your hardware goes smoothly, you might reasonably expect to recoup your investment after a few years of flight operations. But that's a big 'if'. Creating a space vehicle is immensely challenging. Always remember, that's why they call it *rocket science*. If you turn to your investors and guarantee them that your loan repayments will kick in after three or four years, you may get into serious trouble. It would be a disaster if your budding project ran out of cash simply because a three-year development timetable turned into a five-year one and your bankers got cold feet. You must plan for this contingency, because it'll almost certainly happen. Quite apart from your projected passenger-ticket sales or even your hard-and-fast cash deposits, you need an 'added value' element to carry you through any schedule delays, cost overruns or unexpected manufacturing challenges that could frighten the banks.

Option one – and this is a real humdinger if you can pull it off – is to be so wealthy you just don't give a damn what the banks say, because they grovel to *you* for your business, and not the other way round. But the fascinating truth is that none of the top five richest NewSpace entrepreneurs who fit this category are so cavalier with their money, even when they don't have to borrow it. 'Added value' is an essential part of their strategy, because their understanding of that phrase is precisely how they got rich in the first place. Added value basically means boosting that intangible hard-to-quantify-in-terms-of-cash perceived *worth* of your personal image, business reputation or corporate brand.

The faces of Paul Allen and Charles Simonyi are not particularly well known to the public, but Bill Gates wouldn't have been able to build his Microsoft empire without them. Allen was a company co-founder and Simonyi designed Microsoft Word. Their personal fortunes are in the multibillion-dollar range and their added value takes the form of countless philanthropic and charitable interests. Their personal fascination with space is perceived as serious rather than trivial, because of all the other good causes they've been involved in. Allen helped to fund Burt Rutan's SpaceShipOne, while Simonyi made a two-week trip to the International Space Station in April 2007 – and no one ever accused them of wasting money. South African Internet entrepreneur Mark Shuttleworth's 2002 sojourn in orbit was reported by the media in the context of his efforts to improve educational prospects for young people in his home country, along with his development of 'freeware' products for computer users around the world. Iranian-born Anousheh Ansari, a successful business figure and co-founder of the X-Prize, is another orbital enthusiast whose value on the ground is just as great as anything she achieves in space.

And so it should be with your space operation. If you think that building a spacecraft alone will make the project worthwhile, either financially or socially, you are wrong. Privately funded rockets are environmentally and socially controversial luxury items and you must have a solid earthly context into which your dreams of space flight can be neatly plugged. You'll want to generate some additional benefits on the ground, both for yourself and for others. You might, for instance, attract economic development to the dusty desert town where you build your spaceport, or maybe inspire schoolchildren to study science and improve their chances of getting good jobs. Aim for these added-value ambitions, or else you'll just be perceived as some rocket geek of little use to anybody. Sadly, no space business is more likely to go bust before achieving a first flight than one founded by space engineers alone, no matter how brilliant and committed they are. The most perfect spaceship in the world will be grounded forever unless there are some savvy people behind it: people with their feet planted firmly on the ground, even as they turn their gaze to the heavens.

The same added-value principle applies to the giant corporate brands now becoming involved in the NewSpace business, albeit at the behest of their founders. Look at Jeff Bezos's Amazon for starters. For its first half-decade of operation it lost money at a colossal rate, but as a brand it was so successfully marketed that it became famous as one of the few dot.com companies to escape from the crash of the late 1990s relatively unscathed. Amazon's great trick has been to remember that its core Internet bookselling service, although revolutionary-sounding when it was first invented, is only peripherally to do with the Internet itself. Basically its job is to pop books into the post as soon as people have ordered them: a traditional mail-order operation, which the arrival of the Internet has simply made more efficient. In the 1990s a great many dot.com companies were so taken by the glamour of the Internet itself that they slipped up when it came to the dull stuff of actually delivering the real-world services they were offering to customers: sportswear, holidays in the sun, whatever. Bankers were persuaded that the reliability of the Amazon brand was worth more than a few years' unfortunate cash shortfalls and that one day it would come good in raw-profit terms too. Likewise with your space business. Be reliable in everything you do. If you can't afford to deliver on your promises, don't get involved.

Another major NewSpace player is Elon Musk's Paypal company, and again his personal worth is backed up by the perceived importance of his brand. Paypal is *the* globally trusted system for facilitating cash transfers on the Internet securely and reliably, so that ordinary folk as well as big companies can buy and sell stuff. Musk and his team invented PayPal and, once again, amid the virtual hoo-ha of the dot.com devotees, he was satisfying an old-fashioned, real-world need. No one worries about Musk's credit rating when it comes to his new rocket-building company, SpaceX. Amazon and Paypal, then, are brands which deliver the kind of services that are likely to be worth a fortune for decades to come. Bezos and Musk bring added value in spades.

Now we come to the slightly different case of the Virgin Group. Unlike Amazon or Paypal, it's hard to specify any particular aspect of the modern world in which Virgin has pioneered new technologies or marketing styles. Virgin is a globe-spanning corporation that operates by more or less the same kinds of business techniques as any other globe-spanning corporation, investing in markets that it likes the look of, then rebranding them under its own banner. It's into financial services, music, movies, mobile phones, airlines, rail networks and dozens of other activities all geared, in the end, to returning a profit for the group's shareholders. Yet Virgin is perceived as youthful, cuddly and fashionable, and its founder, Richard Branson, brings a personal touch of the adventurer to an otherwise fairly conventional money-making enterprise. His great skill has been to link so many different spheres of business activity under the same brand. If people buy into the logo, they'll buy whatever the logo is painted on. Branson's penchant for high-speed boating adventures, ballooning expeditions and now space tourism is both financed by and adds value to his brand. Think, then, about how you can do the same with your rocket enterprise.

Mind you, the other element that Virgin brings to its suborbital project is a business plan that should work in terms of hard cash. To get the project under way Virgin calculated that it needed to attract at least a thousand potential customers ahead of time before it could commit to building any ships. By 2007 it had processed 1,300 applications and accepted 200 hard-cash deposits. The fleet of five suborbital rockets now under construction (supported by two carrier planes) will cost $150 million by the time they enter service. Then there's an additional $100 million for ground support facilities. A six-seater spaceship marketed to clients at a flat and non-negotiable rate of $200,000 per seat nets $1.2 million gross per flight. Making allowances for running costs and other expenses, Virgin will need to fly perhaps 200 missions to cover its costs. That's 20 flights for each rocket ship: realistically well within the expected airframe lifetime of each one. (The carrier planes that ferry the smaller suborbital space-planes to their release altitude will be able to fly a more or less unlimited number of missions, because the aerodynamic stresses on them are not so great.) Given a flight schedule of two or three times a week, the business should edge into profit after three years – especially taking into account the significant fraction of develop-ment costs already offset by advance ticket sales.

Another trick may be to license copies of the flight hardware to other operators, once Virgin has had a chance to benefit from the first flush of revenues. Burt Rutan's company, Scaled Composites, isn't necessarily geared up to making large production runs. Just turning out that first fleet of five ships will push the company to its limits. But behind the scenes, changes are under way. In 2007 Rutan accepted an offer from the aerospace manufacturing giant Northrop Grumman, on the condition that they left him and his current manage-ment team alone to get on with things in their own way. Grumman accepted the terms, quite literally banking on a future in which more than just half a dozen suborbital rocket ships will be required.

doing business with nasa

Building and flying the ship is the fun bit. Delivering the services you have promised will be the hard part. The story of Kistler Aerospace is sobering. Its proposed K-1 launch vehicle made use of well-known rocket engines already sitting in the hangars of US aerospace companies. The novel element was that the fat cylindrical rocket stages were capped by dome-shaped heat shields capable of surviving re-entry. The whole machine could be recovered by parachute and reused. This wasn't a revolution in space travel so much as a modest incremental improvement, yet that's exactly why Kistler's sales pitch looked so convincing. NASA took note and, as part of its recent Commercial Orbital Transportation Services (COTS) initiative aimed at stimulating private companies to build new ships, it pledged substantial funding, so long as Kistler met certain technical and financial milestones on time. Unfortunately, that didn't happen. We need to look at why this particular dream failed.

The problem with just relying on a governmental space agency as your main client is that you're putting all of your eggs into one notoriously unstable basket. NASA only flies missions occasionally, after many months and years – or even decades – of planning. Typically there might be two or three missions to the Space Station in any year and, just as typically, one of those missions might be deferred or cancelled because of slippages elsewhere in the agency's complex schedule. Robotic missions to Mars go off once every two years (to take advantage of planetary alignments) and other deep-space probes are launched even less frequently. Year on year NASA fights Congress for its annual budget and, as a consequence, multi-year planning for major programmes can become quite fraught. Spacecraft sitting in a hangar half-completed can (and often do) suddenly lose their funding, because Congress have withdrawn support *this* year for a project they loved last year. Either that or NASA will spontaneously revise its internal priorities, responding to changes in the political wind or new faces in the White House.

The COTS scheme is dependent on NASA still being in love with the International Space Station. Actually, that relationship is turning sour and the agency is turning a lustful eye towards its next strategic target, the moon. (More about that later.) There has always been a lot of talk about the Space Station as a 'staging post' to the moon, or even on the way to Mars, but the harsh facts of modern rocketry are this: the Space Station serves no useful purpose in any of the moon/Mars schemes currently under investigation. That's because the hardware for these deep-space missions is lofted away from earth using the same boosters and upper stages that will actually head off to the moon. Likewise, the astronauts leave the surface of the earth in the same craft in which they fly towards the moon. There's no need for any crews or equipment to stop off at the Space Station first. In fact, the rendezvous and docking would simply waste valuable fuel and power reserves. COTS, therefore, is a gamble dependent on NASA's continuing commitment to the Space Station as a worthwhile project in its own right. That cannot be guaranteed and investors courted by the prospective COTS builders, including Kistler, know this only too well.

Your COTS booster will have to find other, private, clients in order to shore up its economic justification. Now the problem shifts gear. None of the private companies, such as TV and mobile phone outfits or earth resource surveyors, or the modestly funded university and academic clients, will have satellites large enough to justify a booster of the size that NASA usually needs. Commercial satellites, as we've already discussed, are quite small and compact. Commercial companies don't build the big scientific probes for interplanetary missions, or the costly human-carrying shuttles and capsules, *except* at the behest of NASA or other government agencies. So who, apart from NASA, is going to purchase the services of your big COTS booster? In October 2007 the Kistler company lost a $278 million contract with NASA because it defaulted on finding the extra $300 million from private investors that it had promised to raise to build its K-1 spacecraft.

But the story isn't so simple. Kistler's annoyed executives wrote to NASA, saying, 'No one is going to invest the capital required to build a vehicle to service the International Space Station if NASA isn't committed to use the capabilities. Our investors were immediately concerned that the assumption underlying our financing may not be as reasonable as originally thought.' It was a classic vicious circle. NASA was unwilling to commit to a spaceship that hadn't flown yet, while Kistler couldn't raise the money to build it unless NASA guaranteed to purchase future flights. The moral of the story is that working with a space agency may not be the best bet if you're a start-up company with nothing else to back you up except a good design for a big booster.

So far, the only COTS contender who doesn't bat an eyelid at these trifling financial worries is Elon Musk, the founder of PayPal,

SpaceX, and the impressive Falcon rocket family that we looked at earlier. As he gladly points out, he doesn't need third-party funding to reach NASA's additional investment requirements. You, on the other hand, might not have such deep pockets.

the euro-tangle

The European Space Agency (ESA) has a funding style that differs from NASA's. Instead of scrabbling year by year for funds, ESA agrees broad mission goals and timetables, then seeks funding for the total extent of each programme, whether it is expected to take two years to complete, or three, or five. ESA ends up with certain agreed budgets to work from and a certain time in which to complete a given project. Everyone knows where they stand. As long as technical milestones are achieved more or less on budget, and on time, there's relatively little likelihood of a sudden nasty cancellation.

But (and it's a big 'but') responsibility for ESA is shared among 17 European countries. A special Council of Ministers sets ESA's goals and, as you can imagine, getting them all to agree on the same priorities is staggeringly difficult. What's more, each country has a different domestic ministry in charge of technology projects, some of which will disagree internally on what they'd like their minister to do. Finally, a maze of aerospace companies gets into the act, lobbying for work in their favoured countries, and so on. The other snag is that if a project does slip behind schedule and over budget the whole negotiation round kicks off again, and can take forever. ESA's decision-making really only works best when things go smoothly.

ESA's biggest mess is its 30-strong fleet of GPS navigation satellites, known collectively as Galileo. More than $2 billion has been spent on this project since it began in 2001, yet so far only one demonstration satellite has flown. Galileo was supposed to end Europe's rather nervous reliance on American GPS satellites, which were funded by the Pentagon. Unfortunately, the Americans were ahead of the game and in May 1996 President Clinton authorised that civilians as well as the military should benefit from accurate signals. The case for Galileo was weakened overnight. Modern in-car navigators and mobile-phone location finders are the latest profitable products of these regulatory changes, but no one wants to *pay* for access to the Galileo signals, because the American GPS data is already available for free. The private investors ESA had been courting to help fund Galileo have therefore melted away. The moral of this story (and the COTS saga too) is *don't* get involved with government space agencies, because the financial and political landscapes they inhabit are liable to change, leaving you in the quicksand without a rope.

If you are an established aerospace manufacturing company already on contract to a space agency, you'll know all the best tricks, including 'cost-plus' accounting, where you charge *whatever* a spacecraft eventually costs to build, and you don't get stuck with paying for all the revisions and redesigns that the space agency has foisted on you at the last minute. Then there's 'fixed price' contracting, which you agree to when you make the deal, because the agency can go to its sponsoring government and say that it has its finances for this project firmly under control, and the hard-pressed taxpayers will get good value for money. Then you wheedle out of the deal some time later, using lawyers who are skilled at that kind of thing. In the last resort you turn to your tame politicians and tell them to tell the space agency that any threat to your favoured space project will damage jobs, etc. Obviously your lawyers will cost more per annum than the budget for most small start-up private space companies, so again, the space agency game is best left to the big players.

However, if you want to shift your chips the other way around and *buy* services, then the Russian Space Agency (RSA) would be glad to listen to your small company's modestly budgeted proposals. If you don't mind a little chaos and have a certain, shall we say, sophistication when it comes to arranging deals with middle-men, fixers and 'special consultants', there are plenty of cool toys to be found in the Russian playroom.

old foes, new allies

In 1992 the US Congress came within one vote of cancelling the International Space Station. There was a lot of grumbling that the multibillion-dollar project had produced nothing but paperwork since its inception by President Reagan almost a decade before. Scared by the prospect of having their biggest project taken away from them, the space cadets reached out to an unlikely friend. On 6 June 1992 NASA chief Dan Goldin talked with his Russian opposite number, Uri Koptev. Their meeting wasn't exactly secret, but neither was it reported in the press. Such caution was understandable. At this early stage, neither man knew what he would find in the other, given the fact that Russia and America had been enemies for four decades. Despite barriers of nationality, history and language, Koptev and Goldin quickly saw how similar their problems were. In those fast-changing times they were both ex-Cold Warriors who had been stripped of power by their political masters and forced into painful spending cuts. They were almost mirror images of each other.

Post-Soviet Russia needed dollars. Within days of Goldin and Koptev's meeting, Presidents George Bush Snr and Boris Yeltsin agreed that NASA would buy access to Russia's Mir space station, and other systems. NASA shuttles would dock with Mir and swap crews around. The Americans also had much to gain. So far, Russian cosmonauts had clocked up 23 years' total flight time between them, while NASA astronauts had accumulated just 8 years. Despite ludicrous economic problems, Russian launches, including all the unmanned probes and satellites, Progress ferries and so forth, had averaged one every five days throughout the 1980s. Alas, the decay set in very rapidly and what we remember from the 1990s is a series of mocking newspaper headlines as the Mir space station began, dangerously, to fall apart.

In July 1991 flight engineer Sergei Krikalev was coming to the end of what he'd expected to be a six-month tour of duty aboard Mir. He conducted a number of spacewalks and worked through a busy schedule of repairs. As October approached he looked forward to the docking of a Soyuz capsule bearing a relief engineer, so that he could go home to his wife and baby daughter. But some of the messages from ground control were making him uneasy. Tactfully, they explained that a special guest astronaut was scheduled to visit Mir in November, but there wasn't enough money for that flight. So they would bring up the guest on the October relief flight, instead of the properly trained engineer they had promised. Krikalev was asked to stay on Mir for another five months.

Krikalev was used to guest astronauts. On his flight up to Mir he had sat next to the astronaut Helen Sharman (the first Briton in space), who had stayed on the station for a few days, subsidised by a London-based consortium. More than a dozen foreigners had visited so far, including a Japanese journalist who spent much of his stay nauseously gripping the walls, as well as serious-minded European astronauts who got on with genuine science programmes. Ticket prices for these visitors ranged from $8 million to $12 million, depending on their length of stay. Recently, it had become almost impossible to schedule any Soyuz launches unless a foreigner was paying for the flight. This was the dawn of the space-tourism industry and it was a shambles, because the RSA was making up its business tactics as it went along.

Mir's latest guest wasn't quite as foreign as Krikalev might have expected, nor would he be paying for the ride. Russia had lost control of its remote launch site at Baikonur in Kazakhstan, because that once-tamed dominion in the south of the Soviet empire had shrugged off the Big Bear's clutches and declared itself an independent country. Kazakhstan demanded that Russia send a Kazakh citizen to Mir, or else Russian space workers would be denied access to Baikonur. And there was more unsettling news. The mission controllers weren't too sure how to tell Krikalev about the tanks that had rumbled through the streets of Moscow that August, during a coup attempt that ultimately failed to reinstate the old Communist regime.

Krikalev's mission badge proclaimed him a cosmonaut of the Union of Soviet Socialist Republics. But down on earth that place or state of mind no longer had any meaning. When he finally came home in March 1992 he had spent 313 days on a little communist-built island adrift in space, while his vast homeland had torn itself to pieces and changed forever. Worst of all, his back salary for that long and exhausting mission was almost worthless, amounting in roubles to maybe a couple of hundred bucks for ten months' work.

In February 1992 Yuri Koptev was given the unenviable task of trying to sort out this mess. He invited American aerospace firms such as Lockheed, Grumman and Boeing to set up offices in Russia and to strike alliances with local factories. These large foreign corporations proved to be valuable allies. They came armed with hard currency instead of useless roubles. Their bargaining power enabled them to impose market disciplines on the Russians – although it was a strange trading environment, which at first left the Americans confused. In the Soviet days factories never worried about profit and loss, or taxes, or much financial control of any kind. They just fudged the figures to fit in with the State's latest five-year plan. Now, with foreigners demanding to know what they might charge for certain goods and services, Russian managers had no idea how to price their products. Western negotiators also found that the notion of a 'binding contract' was alien to Russian business managers, who had seldom come across them in their dealings with the old State.

Meanwhile, NASA and the Russian Space Agency learned how to cooperate, in bad times as well as good. In June 1997 Mir crewmen Aleksander Lazutkin, Vasily Tsibliev and visiting NASA astronaut Michael Foale experienced a terrifying emergency. Tsibliev was using a semi-manual remote-control system to guide an approaching Progress supply ferry towards a docking with the station. It came in much too fast, failed to brake, missed the docking collar entirely and thudded into the side of a laboratory module (Spektr) at sickening speed. The impact destroyed one of the solar arrays and at the same time pushed Mir out of its proper alignment with the sun. Suddenly all of the life support systems' pumps and fans began to fail . . .

The problems piled up. Mir began to spin out of control, turning a complete cartwheel once every six minutes. Tsibliev and Lazutkin clambered into the Soyuz crew ferry and fired its thrusters, trying to stabilise the spin, while Foale, watching the earth's horizon from a small window inside Mir, made an educated guess about the spin rate and called out advice to his Russian companions through the open hatch of the docked Soyuz at the other end of the station. It was a confusing and terrifying experience, caused not by any failure of the crew, but by a series of oversights and misunderstandings on the ground during an earlier redesign of the Progress ferry's guidance system. Foale endeared himself to his Russian colleagues, sharing the workload, helping to save the mission and loyally taking their side in subsequent enquiries. Now *that's* what you call 'international collaboration' in space.

America and Russia today are more or less habituated to orbital collaborations. However, although the Cold War has ended Russia remains unpredictable and still heavily armed. Its economy is still tricky and the temptation to sell its weapons technology abroad is overwhelming. America prefers to keep Russian rocket engineers employed on peaceful international space projects, but the modern Moscow élite is starting to have different ideas as it gains in militaristic confidence once more, backed by a fresh influx of funds from vast oil and gas reserves. Be aware that while Russian space partners will probably always welcome your cash, you might find yourself caught up in the middle of a new Cold War at any moment. If so, then you won't get your money back – and the Americans won't let you play with Russian toys any more. And even if all goes smoothly, you probably won't be able to move a muscle in Moscow without the say-so of some authority in Washington, nervous in case you're thinking of selling Russian rockets to the Libyans or something. Even so, there's a fantastic amount of kit and expertise for you to get your hands on if you're prepared to be patient. And it doesn't necessarily have to cost you millions. If actual launch capability is beyond your purse, you can still offer your clients a taste of the cosmonaut's life, albeit without anyone leaving the ground.

tasting the life of a cosmonaut

In March 1960 the Soviets began constructing a top-secret base hidden deep in a birch and pine forest 25 miles north-east of Moscow. It was called Zvyozdny Gorodok: 'Star City'. A huge square compound was cleared in the middle of the site, thoroughly screened from any nearby roads by the surrounding woodlands. A simple hostel was built, along with standard-issue barracks and a number of low buildings to house the training facilities, some of which (as the cosmonauts would discover to their cost) were designed to inflict stress, trauma, loneliness and exhaustion. In time, Star City would grow to the size of a real town, with its own private community of bars, hotels, sports clubs and administration centres. A short distance to the south, a large and sprawling air-force base at Chkalovksy provided convenient landing strips for jet trainers and cargo planes, as well as accommodation for the pioneering cosmonauts and their young families.

Despite the size of the construction zone, few people outside the space project knew very much about Star City. The road that passes Chkalovsky skirts the dense forest that so effectively conceals the complex. On the right-hand side a small guard post protects an innocent gap in the mask of trees. The turn-off road might easily be mistaken for a loggers' track or an old farm route, except for the guardhouse and the solid tarmac surface capable of taking unusually heavy trucks . . .

Times have changed and Star City is no longer a secret. Today it's one of the largest and most exotic attractions offered by the Virginia-based Space Adventures Company, but also available from several agents worldwide, from around $10,000 a time. Clients dressed in genuine cosmonaut spacesuits train for a day or two inside a Soyuz capsule simulator, and later in a huge swimming pool called a 'neutral-buoyancy tank'. Their spacesuits are fitted with weights that stop them floating to the top of the pool, while not dragging them down to the bottom. They are suspended in the closest approximation to weightlessness that can be achieved without actually flying a plane or a spaceship.

As your clients clamber out of the pool for a much-needed vodka, they will enjoy the *frisson* of the faded Soviet-era décor all around. It's all a bit sad now, but these peeling walls once housed Yuri Gagarin and his kind, back in the long-ago when Russia very nearly ruled the world; and if President Putin's latest pronouncements on space are to be believed, then Star City is due for a fresh lick of paint any day now, as Russia revives a sense of national pride in its rocket achievements. International cash and collaboration on a wide range of space projects will continue to be important to Star City's future, of course. Just so long as the Cold War doesn't get reheated.

building a constituency

Whatever deals you make, whether with space agencies or private investors, you will need to fit your space plans into a broader political and social context. That's always been true for NASA and ESA, and it will be true for you, too, no matter how modest your ambitions.

We often hear that space exploration is a waste of money. Does anyone imagine that dollar bills and euros are rolled up into tight bundles, shoved into the back of the rockets and burned to ashes in the fiery exhaust plumes? If so, then they have not understood the first thing about how the space industry works. Physically a space-craft is just a little speck of metal alloys drifting through the cosmic void. Of the six billion people inhabiting the earth, only a few hundred will have had any direct contact with that ship before launch, and even fewer will communicate with it while it flies; yet the hardware is just one aspect of a project that typically will affect the lives of many thousands of people, and not just in some wishy-washy spiritual way, but by putting food on their tables and paying their mortgages. The impact of rocket technology on the ground is just as important as anything achieved in space.

Quite apart from the obvious aspects of Cold War competition with the Russkies, the 1960s growth of NASA's Apollo project was justified in Congress by the economic benefits it promised for citizens on the ground. The rockets went up into the sky, while the money that paid for them stayed right here on the ground, doing what it was supposed to do. Getting spent. Nothing illustrates this better than the story of why NASA's famous Mission Control centre was *not* built at the obvious place, right alongside the launch towers at Cape Kennedy on the east coast of Florida. Instead it was founded on some barren scrublands in Texas.

We can't help thinking of Texas as the home of big cowboy hats, sprawling cattle ranches and ruthless businessmen playing high-stakes games with their oil empires, aided by glamorous shoulder-padded wives and mistresses. This is a myth today and it was doubly so in the early 1960s. NASA's great political champion Lyndon Johnson may have acted the big man in public, but as a child he had experienced the harshest and most debilitating forms of Texan poverty. Not long after he became John F. Kennedy's Vice President, the British journalist and broadcaster Alistair Cooke received a breathless cablegram from his editors in London: WOULD APPRECIATE ARTICLE ON TEXAS AS BACKGROUND JOHNSON.

COWBOYS, OIL, MILLIONAIRES, HUGE RANCHES, GENERAL CRASS-
NESS, BAD MANNERS ETC. In his subsequent radio broadcast for the
BBC, Cooke lost no time in dispelling these myths. 'Cowboys there
are, west of the Johnson country, which is central Texas, but the
pasture otherwise is so poor that sheep eat the wild flowers and
goats must nibble for the rest. It is very likely that most of the people
here have never seen an oilman. The great oilfields lie to the east and
the north, and to Johnson's family and neighbours they are another
world.'

In 1961 the population living in the treeless, salt-grass coastal
plain alongside Clear Lake, just outside Houston, numbered no more
than 7,000. By 1964 it had jumped to 30,000. There were 5,000
NASA staffers, plus another 5,000 contractors. Apart from the space
professionals, much of the rest of the human expansion was attrib-
utable to Texans eager to take part in the new boom of support
industries, such as house-building for astronauts and technicians,
plus all the catering and cleaning, transport services, secretarial
requirements and myriad everyday jobs created *around* the core
space-control activities. Today, Clear Lake is home to 175,000
people in nine municipalities. Two million tourists visit every year,
many, no doubt, to enjoy the aquatic sporting opportunities of the
lake itself, but most to call on the thrumming heart of NASA's
mission control.

This great Texan social programme also brought considerable
wealth to certain far-sighted business people who quietly bought up
scrublands just outside the mission control compound, knowing full

well that a real-estate expansion was on the cards. Other states benefited too. Hunstville, Alabama was just a sleepy garrison town until NASA sent Wernher von Braun and his rocket team there. Alabamans love space now.

This was what Jim Webb, NASA's chief throughout the 1960s, called 'building a constituency'. He was playing a smart game, ensuring that few politicians would want to block NASA's Apollo project in case they jeopardised space jobs in their home states. Similarly, your best opening gambit when sowing the seeds for your space company would be to contribute campaign funds to a potential senator or president, and keep him or her on your side as they climb the political career ladder. You'll also want to contribute to their political opponents, or else you will be seen as partisan. Appoint a few respected folk from other completely unrelated walks of life to the board of your space company as non-executive directors. They will help you gain the trust of audiences *beyond* those who already believe in what you're doing. You will also want to win plenty of friendly voices in Congress, the most powerful decision-making centre in the world. This you accomplish by offering manufacturing contracts to factories in the representatives' hometowns. Whatever Congress decides to allow in space usually happens. Burt Rutan, Richard Branson, Jeff Bezos and the other major players in suborbital space tourism know that sorting out the politics is the absolute number one priority in any business plan.

spaceport America

The politicians of New Mexico have certainly been persuaded that their state is right for Spaceport America, the projected launch site for Virgin Galactic's fleet of ships, located in southern Sierra County, 45 miles north of Las Cruces. New Mexico boasts calm crystal-clear skies almost every day of the year. Its relatively sparse population ensures a low risk to ground-dwellers from wayward spacecraft, while the skies above the launch site are clear of airliner traffic because of the nearby White Sands missile-testing range. Like 1960s Texas this region of New Mexico is quite poor. State and county officials alike are keen to boost economic activity. Virgin Galactic expects money to flow to the locals: not just from the pockets of its privileged suborbital passengers, but also from the far greater numbers of visitors who will make daytrips to Spaceport America simply to watch the ships take off and land. Many will have driven or flown a long way to make this special visit, and will need overnight accommodation in nearby communities. They'll all need lunch and soft drinks, souvenirs and postcards and so on. Most jobs in and around Spaceport America will be only peripherally to do with actually flying the spacecraft. These added-value components have been an essential part of the 'pitch' for building the place. The only concern is that there might be too rapid an expansion of surrounding infrastructure, detracting from the desert environment but also complicating the launch and landing safety considerations. Certainly no one worries about building Spaceport America in a vacuum, as it were.

From the air, the design for the terminal building and hangar complex looks like a cross between a stingray and a flying saucer, with a hint of the *Millennium Falcon*'s crab-like shape adding a certain populist buzz to the mix. Underneath the exotic and suitably futuristic surface sheen the world-famous architect Norman Foster and his American partners at the URS Corporation have delivered an extremely practical and eco-friendly design that fits, quite literally, into the natural landscape. Much of the roof for the structure will be covered in soils from the surrounding plains. From certain vantage points the building will look like nothing more than a gentle mound rising softly from the terrain, as if it had always been there. Just as significantly, the topsoil cladding serves as an insulating layer. The building's cavernous interiors benefit from stable temperature control with only the barest minimum of energy-hungry air-conditioning.

All of this activity helps to build another constituency, this time among space flight's detractors rather than its supporters. In an age when mass air transport threatens the world by pumping yet more carbon dioxide into the atmosphere and damaging the fragile ozone layer, any space business whose purpose is basically just *fun* needs to think carefully about its environmental impact. The relatively small hybrid rocket engines we encountered earlier don't do a great deal of damage to the sky when flown occasionally, but there are valid concerns about the stresses that thousands of space flights could generate over the coming decades. Virgin's response has been to design their buildings and ground-support facilities so that they are as carbon-neutral as possible. Likewise, the site currently being prepared for the V-2-style Canadian Arrow project is located in coastal Nova Scotia, where the remoteness of the landscape is intimately allied with its environmental sensitivity. So if you're building a spaceport, be sensitive to the concerns of those for whom looking after the earth is a great deal more important than departing it. Build a constituency among the doubters as well as your fans.

climbing the paperwork mountain

The politicians you schmooze will help you deal with the long-standing regulatory bodies involved with aviation and aerospace, some of which have yet to adapt to the growth of private-sector involvement in rocketry. Don't piss off the bureaucrats. Work with them or around them, but never go *against* them. According to an old NASA wisdom, 'Getting to the moon is easy. Doing the paperwork takes time.'

The Federal Aviation Authority (FAA) controls safety over American skies. Without an FAA licence you don't leave the ground. In 2004 the US Congress decided the FAA should also regulate the carrying of passengers into space. The FAA wanted orbital holidaymakers to be as safe in a rocket ship as in a normal plane. No one could guarantee that, so it looked as if space tourism was a nonstarter. But Congress was in a fever to get this glamorous new business moving; and the FAA was also hungry to extend its powers above the atmosphere. In December 2004 Congress ruled that space passengers must simply accept some of the risks. They can't automatically sue you if your flights go wrong. Unless a really rubbish spacecraft kills its passengers or lands on someone's house, the FAA has agreed to relax its rules for the next decade. Your fledgling space businesses will have time to build up a flight history. When your passengers eventually queue up in larger numbers, it will be time to define the safety procedures in more detail.

A bigger bureaucratic hurdle stems from the International Traffic in Arms Regulations (ITAR). These laws guarantee that the simplest sharing of technical expertise between any American technology concern and any another country, no matter how friendly, becomes a nightmare of paperwork and paranoia. Given that you are almost certain to have dealings with a US space operation, aerospace manufacturer, computer provider or federal authority at some point, ITAR may snag you. The American Institute of Physics recently described the regulations as 'inappropriately stifling, a nightmare and probably the single biggest impediment' to international space projects.

Those looking further afield in space will have to untangle the international laws regarding mineral exploitation and land usage on the moon, the asteroids, and Mars. Private industries are keen to see if anything in the lunar soil might be worth a buck or two, but no one is exactly sure what would happen if any company tried to lay claim to a rich patch of acreage. Yet without some clarification in the current laws (dating from 1967) companies might not want to risk investing in the moon. The No.1 extraterrestrial real-estate company *not* to invest in: Lunar Embassy to China, an adventurous outfit that sold 50 acres of moonscape to 34 customers, until it was fined 50,000 yuan ($6,450) by the Beijing authorities in March 2007. You have to be careful with the Beijing authorities. When they tell someone to stop selling plots on the moon, they mean it.

selling to the media

Movie and TV rights may well be the last thing on your mind right now, but don't treat them as last-minute add-ons. While your actual spacefaring clients may number no more than half a dozen at a time, your proxy customers could number in the millions. Provided all of the people aboard your spaceship are willing to participate in a little exploitation (perhaps in return for a share of the revenues) then your enterprise might make a perfect TV drama.

In September 1995 the Hollywood film director James Cameron and his colleague, the renowned underwater photographer Al Giddings, took large-format IMAX cameras to the bottom of the mid-Atlantic, 400 miles south-east of St John's, Newfoundland. Their mission was to photograph the wreck of the *Titanic* in such high quality that the footage could be shown to worldwide cinema audiences. This fantastic adventure was made possible by a pair of Russian deep-sea submersibles, known as Mir. (The Russian word

Mir simply means 'Peace', so it's a popular official name for hardware that makes other countries nervous.) Cameron and Giddings used Mir I as a camera platform, while Mir II carried a rack of floodlights. Floating calmly some 3,800 metres above was a 6,000-ton support ship, *Akademik Keldysh*. Cameron's financial backers were riding on the Hollywood *Titanic* connection, and the prospect of a new feature for the IMAX theatre chain.

Cameron subsequently tried to get the Imax onto Russia's Mir Space Station. He also tried for a seat aboard a NASA shuttle, but politics and launch accidents prevented him from reaching space. He wasn't lucky that time. However, the IMAX gear has flown successfully on several space missions. Astronauts have been trained to use it, and the resulting footage has met with the approval of audiences around the world. The camera can even film in 3D. Footage taken inside the Space Station comes as giddyingly close to being inside it as it's possible for earthbound people to experience. Warner Brothers, the all-encompassing Hollywood media giant, has signed cooperative deals with NASA for further exploits of this kind.

This is something for you to think about. Your short suborbital flights may wear off after a while, in terms of TV and movie appeal, but longer-duration trips to an orbiting habitat or to the moon will make for good soap opera. Likewise, those 'space dives' we encountered earlier. And when a human Mars mission finally gets under way, the media rights could be worth a significant fraction of the mission's overall costs – provided that these rich pickings aren't simply given away by governmental space agencies.

Have you ever wondered why both NASA and its European equivalent ESA simply give all those stunning images and movies to anyone who wants them? The only proviso is that you can't advertise your products or services by using space-agency images. Apart from that if you have a *bona fide* reason for wanting them, most of the images are free. This is because the space agencies believe (and quite rightly, too) that taxpayers have already footed the bills, and deserve a fuss-free return on their investment. Therefore, all the pictures and scientific data that come streaming back from a space mission are openly available. You just have to know where to look for the specific pictures you want and how to ask for the exact data you need. There's so much stuff available it's hard to know where to begin. Your private company, on the other hand, will no doubt pride itself on its refusal to burden the taxpayer. Consequently you get to keep any film and image rights you want. It would be wise to allow the TV and print media access to sexy but generalised shots of your operation, but you may want to keep some of the really good stuff in reserve.

selling to your clients

Most of a typical three-hour suborbital trip will involve climbing to the upper atmosphere and then descending slowly and safely to land again. The actual space phase will only last about five minutes, but that will be enough time for your clients to experience weightlessness and really *know* that they have become astronauts. According to accepted international convention, any person who escapes the atmosphere by travelling to an altitude higher than 60 miles qualifies for that title.

And so far, it's still a gratifyingly exclusive one. In the 50 years of the Space Age, fewer than 500 of the six billion people on earth have entered space, whether for ten minutes aboard early NASA rocket planes, ten days on a trip to the moon or ten months on an orbiting station. The wealth of astronaut missions on the historical record is founded on the skills of repeat visitors: hardened professionals who dedicate most of their adult careers to the job (and mostly on the ground too, undergoing training or helping to train others). Yet despite their myriad technical skills, these old pros won't look down their noses at your greenhorn clients when they sign up for a quick suborbital jaunt. Of all people, career astronauts respect that desire to get up there and see the earth from on high, no matter what the cost or inconvenience. Very few space-agency fliers *like* the fact that so few seats are available for missions, or that the entry requirements for the space club traditionally have been almost impossible for 'ordinary' people to meet.

So the first component of your marketing strategy is to rid your clients of any doubt lingering in their minds that being rich enough to buy a seat on your spaceship won't really make them worthy of the title 'astronaut'. The next important task is to work them pretty damned hard before their actual flight, so they'll feel they've earned the ride once it comes. The knowledge that they are *going* to be worked hard will be an important factor in the layout and writing of your sales literature. When Dr Greg Olsen became the third private entrepreneur to visit the International Space Station in October 2005 he didn't want to be treated like a mere tourist. 'I spent over 900 hours in training,' he said. 'There were many exams, medical and physical, as well as classroom and competency tests. It's not like you pay your money and go on a ride. You have to qualify for this.' As the founder of a New Jersey company specialising in optical and infrared sensors, he conducted serious scientific experiments during his time aboard the Space Station. Dr Olsen took care, however, not to compare his qualifications with those of a full-time space profes- sional. 'I don't consider myself an astronaut. I'm a space traveller and I've been in orbit. But I have far too much respect for astronauts and cosmonauts to call myself that.' Take note that an important and subtle consideration in your business will be the *label* you give to your clients.

As we mentioned earlier, the training regime you offer may well include a basic familiarisation flight aboard a 727 or similar 'vomit comet' jet plane with its cabin completely stripped of chairs, furni- ture and luggage lockers and the inner walls lined with soft padding. The pilot in front will acclimatise your clients to the sudden bursts of acceleration and the sickening turns and stomach-lurching dives that your spacecraft will make. This ride will ensure that when the big day dawns your customers are ready for just about anything. The plane will also fly a series of 'parabolic arcs', accelerating over the crest of an imaginary hill in the sky and then coasting along and dropping down the other side of the arc. For about half a minute or so at a time everyone inside the plane will be able to drift around the cabin and experience a taste of weightlessness. The plane will fly perhaps 10 or 15 arcs before coming home.

It's conceivable that some of your clients will find the vomit comet literally too much to stomach, and will not wish to go ahead with an actual space flight. Be prepared with an 'exit strategy' that leaves them some dignity. It's not a good idea to embarrass rich people. Discreet medical and psychological screening ahead of final ticket sales will help, because if you decide, on health grounds, that some potential clients may not be suitable to fly, you can give them the bad news before they've bragged to their friends that they're about to become astronauts.

The Russian space bureaucrats are masters at this. Several very serious candidates have encountered medical or financial difficulties in signing up for the $20 million rides aboard Soyuz to the International Space Station. At this level of financing and detailed preparation work, and with so very few seats available, a great many things can go wrong and some clients just don't get to fly, despite being eminently qualified. The medical-screening process for your suborbital flights will be substantially less rigorous than for a government space agency's longer-duration orbital missions, but you will need to handle your occasional disappointed customers tactfully.

The suborbital trip itself, of course, has to be more than 'just' a flight, or else your clients might not feel that they've had their money's worth. First impressions count, so ensure that the cabin of your spacecraft is designed with just the same care as the engines, airframe and fuel tanks of your ship. Hire a world-class designer who understands not just what the interior needs to be like, but what people clambering into it for the first time are *hoping* it will be like. In the cultural imagination, spaceship cabins are sleek and futuristic. In real-life, shuttle cockpits, Soyuz cabins and space-station modules are quite shockingly unsexy, and there's a subtle reason for this.

When (for instance) a luxury executive jet is built the production run is large enough that all the interior fittings can be mass-produced to a moulded template. The dashboard of the cockpit has a facade that swishes neatly into the corners to meet the window frames, with air-conditioning vents shaped neatly to match. The TV consoles are stylish, the food trays all fold the same way and the bathroom doesn't have edges that snag. The inside of the passenger compartment looks as if everything fits together into a cohesive and pleasing whole. The same cannot be said of a space-agency spacecraft. These are hand-built machines and sleek styling is absolutely *not* on anyone's priority list. It doesn't make economic sense to create a factory production line for machinery that is only going to be assembled a few times. Remember, there are no mass-production runs here. Consequently there is a nuts-and-bolts look to the interior of real-life spaceships that causes many onlookers to think of them as disappointingly crude, almost wilfully antique.

This is not the dream that your clients are buying into, so give them a cabin that has a certain style. Your first reference point could easily be Stanley Kubrick's epic science-fiction film *2001: A Space Odyssey*. Although it was more than four decades ago when Kubrick's production designers first sat down at their easels and tried to imagine what the spaceships of the year 2001 might look like, the interiors they came up with still look more futuristic than the present-day reality. It's a safe bet that every one of your clients will have seen *2001* when they were kids. Give them plenty of soft padding, a bright and garish colour scheme, some swish control panels with ultra-cool, flat-panel, digital read-outs, and a space meal they can suck through a straw. And give them something exciting to wear: a jumpsuit that looks stylish yet serious, with zippers and special pockets for storing instructions and cameras.

Even if you get all these design touches bang on, you will want to make sure that your customers don't feel let down by the actual physical experience of the flight. Most of them won't settle for staying strapped in their seats. They'll want to float around and look through the windows, but this has to be made possible without them kicking other clients in the face. And there has to be a proper solution to the motion-sickness problem. The smell of someone else's vomit inside the cabin could seriously spoil your clients' experience. You also need to take into consideration their bladder and bowel movements. On smaller suborbital vehicles there simply won't be any room for lavatories, but some passengers may find that, no matter how careful they are to go before the flight, excitement (or fear) loosens up their tummies. Watch what everyone eats and drinks on the last day before a flight, so that no one gets caught short once actually sealed inside the spacecraft.

Finally, you may want to hire some superannuated NASA and ESA astronauts to mingle with your clients on the ground and flatter them a little. Even when their seats are already booked and paid for and your spaceship is standing on the runway warming up for take-off, you need to plan ahead to the moment when you send satisfied clients back into the world as proselytisers for space, and most specifically, for *your* particular method of reaching space. After all, rich people have rich friends, and you want them to be your next customers.

persuading everyone else

Many people are fascinated by space exploration and plenty more take at least a passing interest, but those multitudes are probably not sufficient to ensure a long-term continuation of the human space programme at a major national level. The drama of rocket flight is as great now as it ever was, but the *perception* of it is not coming across to a global audience in the way that it used to. Astronaut-heroes no longer seem to inspire whole societies as they once could, and rockets have lost much of their ability to inspire awe. There is a problem in the language of space if young people no longer find so much excitement in the adventure today as their parents did when they were young. Perhaps NASA and its international partners need, after all, to speak to the younger generation in a more poetic way that engages our emotions as well as our minds. Perhaps your private company can reconnect ordinary people emotionally to the drama of space travel.

If so, one of the themes you'll need to address is the danger. So far, as we touched on earlier, approximately 500 people have flown into space, sometimes just once, many repeatedly. Of those 500, three were killed before their spacecraft ever left the earth; another crashed to earth like a meteorite after a hectic uncontrolled re-entry; three were suffocated in space just before their homeward-bound capsule reached the pressurised haven of the atmosphere; seven were blown apart moments after launch; and seven more died in a high-altitude disintegration of their ship. This, then, is a fatality rate of one space traveller out of every twenty-five. A dangerous profession indeed. Yet somehow the illusion persists that rocket flights have become routine and that accidents are somehow unusual – and blameworthy.

Even the so-called 'flawless' space missions are fraught with blood-chilling incident. We all remember (with a little help from Tom Hanks's movie) that Apollo 13 blew up on the way to the moon and that its crew only narrowly avoided being stranded forever in deep space, yet how many people know that Apollo 11's lunar module *Eagle* came within seconds of crashing to pieces as it headed towards that triumphant first touchdown of July 1969?

In the TV bulletins Buzz Aldrin's static-encrusted voice said something like '1201'. Then the Capsule Communicator (CAPCOM), the astronaut entrusted to be the voice of the crowded Mission Control room in Houston, said, 'Copy.' A crackly pause, then '1202' from Aldrin. 'Um, you're go on the 1202,' said CAPCOM. For several minutes after that, no one spoke. Then CAPCOM cut in: 'Sixty seconds.' At last, Neil Armstrong's voice came on the loop in more confident vein: 'Houston, Tranquility base here. The *Eagle* has landed.' Then, almost drowned out by the applause of other mission controllers, came the little comment from CAPCOM to his buddies aboard Apollo that most news pundits and TV audiences missed: 'You've got a bunch of guys about to turn blue. We're breathing again. Thanks a lot.'

In Mission Control they'd all been holding their breaths, wracked with tension. It's hard for outsiders really to grasp the frightening significance of the clipped dialogue and the astronauts' shorthand way of talking about the problems that had accumulated inside the lunar module's cabin. '1201' and '1202' were numerical displays from the fragile little computer, warning that it was about to over-load. Taking one hell of a gamble, a young digital expert at Mission Control told CAPCOM to relay to Apollo his verdict that the warning could be ignored. The truth was, he didn't know for sure. If he had been wrong, *Eagle* would have spun out of control. Nothing more than gut instinct and a vague recollection of similar spurious signals cropping up during rehearsals on the ground guided his decision. And later, when CAPCOM said 'Sixty seconds', he really meant, 'Hey guys, I don't want to worry you, but according to my read-outs, you've only got about a minute's worth of fuel left before you crash.' In fact, Armstrong was trying to avoid a very large deep-walled crater ('big enough to house the Houston Astrodome,' he recalled later) and it took him almost all his remaining fuel to find a safe and level place to land.

To a busy astronaut absorbing data from myriad control panels while trying to read the runes of an alien lunar landscape twisting and turning outside the cockpit window, too much babble from Mission Control wouldn't have been half so welcome as the quick, simple reminder, 'Sixty seconds.' But to the public, and even the supposedly knowledgeable journalists listening in as these events unfolded, such terse exchanges were a barrier to understanding. As the writer Norman Mailer confessed at the time, he didn't grasp at all what was happening: 'I was bored. Sitting in the Mission Control press gallery, I noticed the other reporters were also bored. We all knew the engine burn would succeed and Apollo 11 would make the proper descent. There seemed no question of failure. I could not forgive the astronauts their resolute avoidance of an heroic posture.'

This unavoidable gulf between NASA-speak and the rest of the world remains a problem today, for people are still surprised when missions go wrong. No language has yet been invented that can persuade the public to fund a basically very *dangerous* enterprise. They expect that NASA should be able to predict risks and that somehow space vehicles fuelled with the energy equivalent of small thermonuclear bombs should be made as safe to fly as ordinary aircraft; at least, they should be if the Government is paying for them with tax dollars. Surely all that dry technical babble means that NASA *must* understand what it's doing by now, so any failures have to be the result of incompetence?

But rockets are dangerous and outer space will always be a tricky environment. People make errors of judgement, sometimes very serious ones that claim lives. Equally, someone's gut instincts can save a mission, even when all the technical instruments and read-outs spell trouble. Apollo 11 *could* so easily have crashed, plunging the United States into a fantastic international humiliation that would have echoed to this day. The shuttle *Columbia* could so easily *not* have been struck by fuel tank debris in January 2003 . . . Once again, history has something to tell you about how to present your private space company to the world. The white-knuckle drama of space flight *is* the truth. The 'routine' is the lie. You might want to find a way to improve on NASA's stilted, unemotional PR-style, even if your cockpit chatter remains terse for practical reasons.

A legal nicety: the FAA might have waived some of its intrusive regulatory powers for the next few years, allowing you to build up a flight history and a set of safety statistics based on reality rather than wishful thinking, and you may well persuade your clients that if they fly with you, they are essentially risk-taking pioneers who must accept that their journey into space will not necessarily be as safe as they would expect from an average airliner trip. However, the *families* of these people do not sign any waivers and they may sue you if your vehicle encounters problems or has 'a bad day', as they say in the rocket business. Angry relatives might think you put their loved ones into a spacecraft without fully informing them of the potential hazards. So don't pretend that what you're doing is 'safe' in the way that most people understand that word or the lawyers will shut you down at the first sign of trouble.

five   islands in the sky

If your ambitions stretch beyond making brief suborbital hops and
you want to reach into full orbit, it's a good idea to ask yourself why.
At $100 billion the International Space Station is the costliest and
most complicated engineering project in the history of the world and
after some two decades of cost overruns, launch delays, shuttle accidents and countless political arguments, it hasn't delivered much by
way of tangible scientific benefits. Even NASA falls short of making
such a claim. But the Space Station dream refuses to fade.

In 1869 a progressively minded American author, Edward Everett

Hale, wrote a story called *The Brick Moon*. A massive spherical assemblage of bricks 200 feet in diameter is catapulted into space between two giant rotating flywheels. It's supposed to be a beacon, a reflective blip in the sky that can be observed by ocean-going ships and used as a reference for navigation and timekeeping: a sort of Victorian forerunner of GPS SatNav. The catch in Hale's story is that some of the sphere's builders accidentally become passengers when it is thrown aloft. They communicate their plight using mirror semaphore. Back on earth the flywheels are powered up again to send food and supplies up to the stranded workers. And that, ladies and gentlemen, is the first description of anything like a 'space station' on record.

By the 1920s astronautical theory was advancing at a great pace. Rocket pioneers such as Robert Goddard (who invented that essential LOX/kerosone combination we've already encountered) in America or Hermann Oberth and Willy Ley in Germany understood the potential for space stations, but the first thoroughly detailed engineering proposal appeared in Hermann Noordung's *The Problem of Space Travel: The Rocket Motor* (1927). Very little of Noordung's personal background is known today. His real name was Herman Potočnik and he was an unremarkable officer in the Austrian Imperial Army who died in Vienna in 1936, poverty-stricken and tragically young, after catching pneumonia. Not much of a life, but his ideas are prescient even now, at the dawn of the twenty-first century. He described a *Wohnrad* or 'Living Wheel' whose gentle rotation would provide its crew with artificial gravity. Airlocks and safety bulkheads were all taken into consideration, along with a huge parabolic dish that would collect and focus sunlight for power. The docking airlock rotated at the same rate as the station, but in the opposite direction, maintaining its position relative to the horizon. Rocket ships could approach without themselves having to tumble. Noordung's station was supposed to be an astronomical observatory and a navigation beacon: that whole GPS thing again.

In the early 1940s the British futurologist and science-fiction writer Arthur C. Clarke helped develop radio navigation systems for allied aircraft returning from their bombing raids over Germany. At the end of the war he recognised that his cherished dreams of space flight would never be achieved unless governments and industrial investors could be persuaded to see its economic benefits. In October 1945 he published an article in *Wireless World*, an electronics magazine, essentially outlining the modern concept of geosynchronous communications satellites (we've talked about that special orbital realm before). The only difference between Clarke's ideas of 1945 and the global network we see today is that he was thinking in terms of bulky and unreliable 1940s radio equipment, all glowing thermionic valves and clattering circuit-breakers and wiring boards like giant bags of knitting. He suggested that his 'Extraterrestrial Relays' would have to be staffed by human engineers, recruited in all probability from that mighty organ of 1940s telecoms advancement, the British Post Office. Clarke's outpost had a radio room, a small library, a kitchen, a surgery, cabins for the crew and, of course, separate and more luxurious quarters for the station's commander. It was a very English space station. The swift development of microchip technology put paid to that idea, eliminating any need for Post Office sparkies in orbit.

Next up on the space cadets' wish list, especially once the real-life rocket age began, was an even more giant space station capable of harvesting solar power. The collected energy is converted into a microwave beam which is then fired down to earth. On the ground, microwave dishes capture the beam and convert it into usable electricity. The advantage of this idea is that the sun's power is free and limitless. The technology for a space power station has existed for many decades. The disadvantages, however, are daunting. To make any significant contributions to earth's electrical hunger, a space power station would need several square miles of receiver panels. If you object to wind farms spoiling the view, you'll hate this idea too. Another possibility is to concentrate the beam ferociously onto just one small ground receiver. Now the problem becomes, what happens if the beam strays from its target? It would cook anything in its path. Imagine some supervillain taking control of it and threatening the world with an appallingly clichéd movie plot . . . The solar experts' frustrations are understandable. It is a real pity that we can't make better use of the sun to produce electricity on the ground, but so far, unfortunately, no one has come up with the right balance of cost and environmental impact to justify building a solar space station.

Then there were the economic and military potentials of space stations. In May 1963 NASA astronaut Gordon Cooper, hunched inside his cramped Mercury capsule, amazed mission controllers when he reported being able to see roads, buildings and even smoke from chimneys. By 1964 the clarity of the earth as seen from space could no longer be denied. Astronauts in two-man Gemini capsules corroborated each other's impressions and returned to earth with colour photographs of drainage channels, wheat fields and some interesting airfields and other man-made structures in eastern Europe and Russia. One photo also revealed with startling clarity the differences in colour between adjacent agricultural plots in Texas, some of which had received sufficient water to deliver healthy crops, and others where the crops were drier. Down on the ground, the farmers couldn't spot the subtle colour differences between a good plant and a slightly poorer one, but from up in space the cumulative differences were clearer. Obviously observations like these had great economic significance.

Such, at any rate, were the kinds of exciting discoveries made by NASA and the civilian space effort in the first decade of the Space Age, and announced to the public. Behind the scenes, a vast industry, funded just as lavishly as the human space programme, was functioning in conditions of such secrecy that most politicians in Washington weren't permitted to know about it. If your company wants to develop spy satellite technology for the Department of Defense, the CIA or the National Reconnaissance Office, then you might not want to tell anyone. And don't try to sell them your space-station ideas. You'll find that these agencies long ago abandoned interest in them, because they don't like the idea of astronauts jiggling around and setting up vibrations that disturb the cameras. The orbital spy trade is for robots only.

worldships and space colonies

So now we're into the 1970s and the first tentative steps towards real space stations: Russia's bus-sized Salyut modules and NASA's roomier Skylab, carried into orbit in 1973 atop the last of its giant Saturn V rockets. These missions were aimed at seeing whether or not humans could live safely and sanely in space for long periods. As we'll discuss later, the results were mixed, though not a deal-breaker.

Meanwhile Gerard K. O'Neill at Princeton University proposed huge rotating cylindrical colonies in space as a way of easing the population pressures on earth. The structures would be perhaps three miles long and fabricated from materials processed in lunar factories. These weren't new ideas. In 1903 the great Russian pioneer of astronautics Konstantin Tsiolkovsky (who worked on detailed ideas about rockets from the 1880s until his death in 1935) proposed a huge habitable cylinder, spinning on its axis and containing a greenhouse with a self-supporting ecological system; while in his novel *The World, the Flesh and the Devil* (1929) J. D. Bernal devised Bernal Spheres: self-supporting 'Worldships' capable of housing many thousands of inhabitants. O'Neill's colonies were intended as a theoretical exercise to stretch the imaginations of his physics students, but the scheme hit a nerve with its detailed studies of closed ecological systems: an urgent fascination for environmental campaigners, because the earth itself is just such a system. Especially popular was O'Neill's proposal that all heavy industry and energy production should be taken into space, thus saving the ground from the burdens of pollution.

The fly in the ointment was that NASA's emerging space shuttle technology simply wasn't the do-it-all space truck everyone had hoped it might be. O'Neill's people couldn't get their pieces into orbit cheaply enough for the scheme to work. And when the environmentalists who had at first heralded the proposal got to see how many rocket launches would be needed to kick-start a colony, they had second thoughts about the whole thing. Giant colonies were put on the back-burner, and yet another space station dream was shattered.

Like so many space visionaries before them, O'Neill and his supporters hitched their wagons to a compelling American dream: the frontier spirit of the Old West. They adapted the country's romantic self-image: the voyage of Columbus, the Mayflower and its Founding Fathers, the pioneering families on the Oregon Trail, all that stuff. History seemed to offer a compelling model. A small fleet of ships sets sail across an uncharted ocean to discover new territory and a 'colony' is established at the point of landing. (Best not to dwell on the settlers killing any indigenous populations they encounter.) The colony becomes a thriving, permanent town. In time other ships arrive carrying families to populate the new territories opening up inland. Inspired by Wernher von Braun and the earlier triumphs of the Apollo missions to the moon, space visionaries talked in terms of 'conquering the sun's empire' and 'colonising the space frontier'. Always, their plans were worked out with exact attention to fuel weights, rocket thrusts, orbital heights and speeds. As Arthur C. Clarke has pointed out, 'No achievement in human affairs was ever so well documented before the fact as space travel.'

The flaw in the plan was simple. Where was the trade? Those colonies could be sent into space, but what could they be expected to bring back as a return on the huge investment of building them? This fundamental economic problem could not be wished away by any amount of romantic language or by vacuous dreams of harnessing solar power. By the 1980s, with the Cold War Space Race long since ended, there seemed no obvious way of justifying large-scale astronaut space programmes of any kind. Always remember, for your space business you can't just sell a dream to your investors. You have to deliver them a cash profit too.

The second problem has been even more of a bummer for potential space colonists. What kind of a spaceship would haul the equipment into orbit? It would be a good few years before the colony business could rely solely on building materials dug out of the moon. Most of the visionaries who try to sell us on these grand schemes tend to skip forward in their PowerPoint presentations to the juicy years far in the future, steering clear of the difficult first decade. The technical designs for a zillion space habitats have been detailed down to the last nut and bolt, but the delivery trucks for all the girders, wall panels and fitted carpets have never been specified, unless you count optimistic airbrushed 1970s illustrations of NASA space shuttles arriving in orbit two or three at a time, twice a month, their cargo bays brimming with goodies. In the end, NASA turned out not to be the organisation best suited to delivering this vision, so the dream of colonising space faded away.

Now a startling new idea has emerged. Instead of thinking on an unrealistically grand scale, why not build small space stations, just for *fun*? It would require only a few dozen rich holidaymakers to make a boutique orbiting hotel worthwhile.

that floating feeling

Where once the gentle spinning of a space station was seen as a way to make life more or less bearable for its crew, total weightlessness is now regarded as the main justification for visiting one at all. This is the ultimate explanation for why Skylab, Mir, and now the International Space Station look nothing like the speculative visions of the early space theorists. They can be as haphazard and inelegantly shaped as they like, because they don't have to spin. As for your private space-tourism company, it's comforting to know that weightlessness is one of the fundamental thrills you want to deliver, so you don't have to worry about spinning your space station like a giant Catherine wheel, lamely trying to create a gravitational echo of life on earth. Nor does your space station have to be particularly large, because your clients (the private ones at least) won't be living aboard it for more than a few days or weeks at a time.

The biggest selling point for your station will be a set of large 'viewing ports', backed by the promise of plenty of time to gaze through them at the earth below. Of all the frustrations experienced by professional astronauts, the one that upsets them most is how little time they are able to spend just looking out of the window during a busy mission. Make sure your clients don't feel guilty for just . . . looking.

living room

NASA's public relations team are fond of explaining that the International Space Station is the size of two football fields. What they aren't so keen to point out is that the living and working compartments inside this labyrinthine leviathan are barely more spacious than the interior of a bus – and a bus packed from floor to ceiling with luggage at that. True, the modules as a whole deliver an interior volume equivalent to a pair of 747 jumbo jets, but this is the product of adding up the cramped spaces inside a jumble of small pieces, essentially all shaped like narrow corridors. The uncomfortable truth is that because of the limited diameter of Russian and

American launch vehicles, no *single* module can be more than 15 feet wide. The Space Station's modules are really just cylindrical tubes crammed full of lockers and equipment. There's scarcely room to swing the proverbial cat.

But modules don't necessarily have to be this cramped. A decade ago, Donna Fender, a project manager at NASA, came up with a radical scheme. She started out with a simple challenge, based on two problems. 'First, at $10,000 per pound launch costs, you want to get your spacecraft as lightweight as possible. Second, you want it as large as possible, especially when people have to live inside it for months or years at a time. Normal aluminium modules don't seem like the answer. We had a better idea. Don't build 'em. Blow 'em up!'

The first inflatable object in space was the Echo 1 balloon, launched in 1960 to bounce radio signals back to earth. It expanded to 100 feet in diameter and the sunlight glinting off its shiny surface made it clearly visible in the night sky, but in the end it was just a simple hollow ball. Fender and her people put a new spin on an old idea and came up with the Transit Habitation Module (TransHab). Made from flexible fabrics instead of metal, it fits snugly into a rocket cargo bay for launch, then expands like a balloon to double its volume once deployed in orbit. The fabric can be folded up inside the rocket, but it doesn't stretch like rubber. It is resistant against meteorite impacts and forms a rigid shell after inflation. Foldaway floors then lock into position, providing instant accommodation for six people. TransHab delivers two and a half times the room of an aluminium module and at less than five tons for the main shell it's half the weight.

The biggest threat to any balloon is a pin travelling with sufficient force. Deep under Cheyenne Mountain in the American Rockies, the Space Surveillance Center keeps radar tabs on more than 8,000 orbiting targets, ranging in size from discarded rocket stages to lost spanners. Only five per cent of those things are functioning space satellites. The rest is a swarm of dead junk, 3,000 tons of metal and plastic swirling about at full-on orbital speeds. But TransHab's designers aren't too concerned. Their secret is Kevlar, a bullet-proof fabric woven from flexible but unbreakable carbon filaments. TransHab's skin is constructed from three layers of Kevlar interleaved with airtight neoprene. The outermost cladding is lightweight Nextel foam. In contrast to the Kevlar, Nextel is as easy to tear as sponge cake, but strength isn't its main purpose. It is designed to vaporise into hot gas the instant a rogue particle hits. This dissipates the impact energy in all directions and blunts a particle before it can punch a hole all the way through to TransHab's interior.

In 1998 the first prototype TransHab was installed at the bottom of a huge water tank at NASA's Johnson Space Flight Center and inflated to double the internal air pressure planned for normal operations. And it didn't surrender so much as a single bubble of the air inside it. Another test at four times operating pressure also passed off flawlessly. The final challenge was to install the module inside a huge vacuum chamber, a relic from the Apollo lunar-landing project, used in the 1960s to test spacecraft for leaks. It was as close as the designers could get, for now, to going into space without actually leaving the ground. Again, everything worked correctly. And TransHab wasn't complicated to build. All the fabric panels for the first prototype took the designers just three weeks to stitch together by hand.

At 25 feet in diameter and 23 feet tall TransHab would make a tempting starter home here on earth, never mind in space. The uppermost floor is an exercise area, complete with a treadmill, a medical station, a changing area and a shower cubicle. The second level incorporates six separate private bedrooms, along with life support and air-conditioning equipment. The bedrooms provide each occupant with about 80 cubic feet of private accommodation, with room for a desk, a computer and even a personal entertainment system. The entire sleeping area is double-walled. The two-inch gap between the walls is filled with water, which provides protection against radiation and also blocks off noise, helping everyone get a good night's sleep.

The third level provides a social gathering space with seating for up to a dozen people at a time around a large dining table. An ultra-strong but lightweight carbon composite core runs from top to bottom of TransHab and there's a corridor that the occupants can use to float from one floor to another. A docking airlock at the top, with a double hatch, allows for connection with other modules. For extra strength, carbon-reinforced ribs are woven into TransHab's outer fabric. Stored for launch in the rocket's nose cone, the whole assembly packs up tight, with the strengthening rods and central tunnel lying alongside each other like arrows in a quiver. The effect after inflation is pretty amazing.

Financially, everything was going quite well until Fender and her team tried to find a role for their cherished project as a module for the International Space Station. NASA's political enemies in Congress refused to allow TransHab any more government funds. The major aerospace companies responsible for building the Space Station's clunky metal modules were not exactly thrilled about bright new ideas from competitors. The corporate spin doctors worked the corridors of Washington, protecting the old expensive metallic systems. Meanwhile, delays and the excessive costs of the Space Station alarmed even the most supportive politicians.

TransHab stimulated a lively debate about NASA's future, even among its supporters. Was clumsy government money really the best way to fund space exploration? Some Space Station components had been sitting on the ground throughout most of the 1990s waiting for launch. In that time privately funded technologies such as the Internet had flourished; 'intelligent' materials had been invented that made aluminium modules look like tin cans; and the TransHab team had reinvented space-station technology in just three years. Yet NASA as a whole seemed unable to respond to these rapid changes. The Space Station was looking dangerously old-fashioned even as it got into its stride. But Fender wasn't going to let her idea fade away. She made a bold announcement in the *Wall Street Journal*, telling investors to come and take a look at TransHab.

Enter Las Vegas businessman Robert Bigelow, banking entrepreneur, creator of the Budget Suites of America hotel empire and dedicated space visionary. He licensed the TransHab design from NASA and in 2006 successfully tested an unmanned prototype in orbit with help from a Russian Dniepr booster rocket. A second and larger test module flew just as successfully a few months later. Bigelow expects to offer four-week stays in orbit aboard his Sundancer module for $12 million, barely half the price on offer from the Russian partners in the International Space Station. Governments and corporations can also lease Sundancer, in part or in whole, for research work conducted by crews of up to four astronauts. A full year's privileged access will set them back $88 million: a hefty sum, but still a fraction of the running costs associated with the Space Station.

Bigelow is the first person to admit that 'You can't count on any business model that is too dependent on NASA.' Spreading his risks and widening his potential market, he is promoting Sundancer to 50 different countries, many of which will welcome the chance to fly their own astronauts without having to invest in an entire space infrastructure. Bigelow's other risk-reduction strategy has to do with his financial partners. He costs his project using the jargon and techniques of the real-estate business: a language that Wall Street can understand. All he's doing is selling space in space rather than on the ground.

Sundancer was originally supposed to fly in 2012, but the success of the unmanned test modules has brought that date forward to 2010 or thereabouts. Everything will depend on the courtesy shuttle that delivers customers to Sundancer and brings them home again. Development of a genuinely reusable orbital transporter remains *the* last great unconquered space frontier for our generation. As we mentioned earlier, Bigelow is taking the X Prize idea one step further by offering a $50 million reward for the first company to go all the way into orbit with a passenger ship.

soyuz for sale

Meanwhile, the trusty Russian Soyuz capsule could play a role. There's a popular misconception that Russian space technology is a bit primitive. Certainly it's simpler than NASA's, but it works more reliably. The International Space Station is dependent on Russian Soyuz capsules and their uncrewed cousins, Progress cargo ferries. The US government has recently made several deals with the Russian Space Agency (RSA) to pay for Soyuz access. The Soyuz is also the only 'lifeboat' if anything goes wrong aboard the Space Station.

The Soyuz's first flight on 23 April 1967 was a disaster. Solo cosmonaut Vladimir Komarov was killed when the landing parachutes failed and his capsule smashed into the ground like a meteorite. Subsequent flights went more smoothly, but on 29 June 1971 Georgy Dobrovolsky, Vladislav Volkov and Viktor Patsayev suffocated when the air leaked out of their cabin as they tried to come home after a trip to Russia's first space station, Salyut 1. The little ship's record since then has been impressive and although the electronics have been updated the basic exterior design has hardly changed in four decades. It has flown more than a hundred missions in the last 35 years with no fatalities. There have been a few hair-raising close calls – including a launch-pad abort when the booster rocket exploded and another abort at high altitude – but these incidents highlight just how safe the capsule really is. Escape rockets pulled the cosmonauts safely away from disaster and all lived to tell the tale. To put it bluntly, four cosmonauts have died in the course of 120 Soyuz missions over 40 years, while the supposedly more sophisticated space shuttle has lost 14 astronauts in 120 flights across 25 years.

For all its reliability, the Soyuz can't solve all of your orbital transport problems. It carries just three people, and two of them have to be Russian pilots. It's not reusable, the landings are bumpy and there's minimal control once the crew compartment hits the atmosphere, so the capsules have to come down in remote, unpopulated areas.

The Soyuz's builders at the RSC-Energia Corporation think they've come up with a better idea. At the 2004 Paris Air Show in Le Bourget a full-scale model of a new spaceship was unveiled in a bid to attract foreign investment. The Kliper ('Clipper') will build on the strengths of the Soyuz while adding a few extras: three more seats in the cabin, fully reusable core components, and a pair of stubby wings for a smoother and more controllable homecoming. If the ESA and Japan sign up for the Kliper, their astronauts will have a guaranteed ride into space, instead of having to bargain for occasional spare seats on a Soyuz or wait for NASA's Space Shuttle to be repaired or replaced. The Soyuz has won considerable respect in Europe. Its successor could prove even more tempting – and better value for money – than anything NASA can offer. The Russian government is expected to invest heavily in Kliper, although foreign cash will still be essential. Perhaps you can be an investor, especially if you have some finesse for handling Russian politics and its often rather tangled business dealings.

There's nothing too complex aboard Kliper. A removable aerodynamic shell fits around a pressurised cabin with a basic cylindrical design. At the rear a familiar plum-shaped docking module (adapted from the front of a Soyuz) enables docking with the Space Station's existing hatches. Kliper will also take over the unmanned cargo role of the Progress ferries and on these occasions the aerodynamic shell will be left behind to save weight, while a simplified version of the core module is packed with supplies and sent aloft. What looks on the outside like a brand-new spacecraft is actually a clever and more stylishly streamlined update of available Russian technology.

Manned flights of the Kliper are likely to be launched aboard a revised version of the 50-year-old Soyuz rocket, almost identical to the ones that launched *Sputnik* and Yuri Gagarin into orbit. Again, the emphasis is on adapting well-understood designs from the past. There is a penalty, however. A Soyuz spacecraft weighs eight tons; the Kliper weighs thirteen, which is beyond the lifting capacity of

Soyuz's current booster. So the new ship might fly, instead, on a Zenit rocket built in the Ukraine; or perhaps an Angara-A3, a new booster currently under development as a commercial satellite launcher. Everything depends on how much money Europe, Japan and other countries are willing to contribute. Another and perhaps simpler option is to stick with the old Soyuz booster, but build one small additional stage to give the Kliper a final little kick into orbit.

chinese puzzles

A recent and very belated Chinese entry into the 'space race' may rack up international tensions in orbit. China's Shenzhou spacecraft is similar in size and shape to Soyuz, from which most of its best ideas have been 'borrowed'. The rear of the craft is a service module containing propulsion and life-support systems and sprouting a pair of solar panels to boost the electricity supply. The middle section is the main crew capsule. It has a heat shield at the rear, parachutes to slow its final descent through the atmosphere and small retro-rockets to lessen the shock of touchdown, while the front end is a disposable docking module that allows the Soyuz to latch onto space stations.

As far as western experts can tell, Shinzou is a close copy of its Soyuz equivalent, except that its front end is nearly twice the size and has a life of its own with thruster rockets, solar panels and guidance systems entirely independent of the rest of the ship. The front modules always continue flying long after the crew capsule has returned to earth. These mysterious modules are designed to be placed into orbit and left there, just like miniature space-station components. At the moment they fly on their own, but they have hatches that can link together. It would be a surprise if China didn't build a space station eventually.

Shenzhou and a wide variety of other projects are in the hands of the Chinese Aerospace and Science Corporation (CASC). It controls 300 organisations around that vast country and employs nearly 250,000 scientists and engineers. Many of the lower-level workers are drawn from the Chinese People's Liberation Army. The cost of these efforts is staggering for a developing country, even one with a decades-long record of Asian Tiger economic growth. Western experts estimate China's annual spending on space at $3 billion. While that's no more than one-fifth of NASA's budget, it's ten times more than Russia can currently afford. Whether or not the Chinese plan to let any potential Western users near their hardware is anybody's guess, but you might want to keep an eye on them and see how their inscrutable rocket programme affects the political and financial mood of the entire international space game, for better or for worse.

the human factor

Whatever your plans for the orbital realm or beyond, the most unpredictable components involved in your schemes won't lie within the motherboards and heat shields or the engines and turbo pumps. For all the volatility of your hardware, the trickiest elements to deal with will be the brains of your people.

Imagine the following scenario. An interplanetary spacecraft is six months out from earth on a historic mission. Everyone aboard has been selected with great care and no one could have predicted what comes next. During a live broadcast to earth a vicious fight breaks out among the crew for the whole world to see. Mission Control tries to calm the situation, but suddenly the radio and picture links go dead and all contact with the spaceship is lost . . . This is the nightmare scenario facing NASA and its partners as they prepare to build a lunar colony by the year 2020, with the possibility of a human expedition to Mars in 2032. The evidence that even the best crews can go mad when cooped up together for many months at a time is already available here on earth.

During the Antarctic winter the sun dips below the horizon and doesn't rise again for six months. A 1996 diary entry from Antarctic researcher Chris Bero shows the strain of spending long periods trapped indoors because of the harsh conditions outside: 'You can watch several thousand videos. You can run on a treadmill, lift weights, play volleyball, slaughter your co-workers in a virtual-reality game, or you can just sit and watch the same people day in, day out and estimate the rate of growth for their facial hair.' These conditions closely match those of a long space mission. It's permanently dark outside your window, your living quarters are cramped, you can't go out unless you wrap up in a cumbersome protective suit, and you can't escape from people you find annoying. At one point Bero fantasised about spraying one of his colleagues with acidic chemicals and burying him in a snowdrift.

A more serious Antarctic incident from 50 years ago haunts the NASA experts in charge of astronauts' mental health. In 1957 a construction worker at an Antarctic base lost his mind and became violent. His colleagues prepared a special room for him lined with mattresses, so he wouldn't injure himself bashing against the walls. The man spent an entire winter locked away. What would happen if someone lost their sanity on one of your spaceships or orbiting hotels?

NASA's thousand-page instruction manual for coping with medical emergencies on the International Space Station covers everything from toothache to broken bones and stomach problems. Five pages deal with possible mental breakdowns: 'Restrain the patient using tape around wrists, ankles, and use a bungee around the torso. If necessary, restrain the head, place a rolled towel under the neck and restrain with tape.' It is then up to the Station's commander to decide if the 'patient' should be sedated and sent back to earth. This would require a special shuttle flight or, more likely, bringing forward the date for a Soyuz capsule re-entry. There's always a Soyuz plugged onto the Station ready to act as a lifeboat in case something like this happens. Astronauts on the Space Station always know that earth is comfortingly close. However, on a long mission to Mars there will be no possibility of coming home at short notice if the strain gets too much.

artificial insanity

The Institute of Medical and Biological Problems (IMBP) in Moscow has been around since the dawn of the Space Age. All of Russia's cosmonauts have to pass the IMBP's strict medical tests, but not everyone who passes through this rather scary building is destined to reach orbit. Non-cosmonaut volunteers sign up to go into 'isolation chambers', basically pretending to go on long space missions without actually leaving the ground. Their only contact with the outside world is through a microphone link and they are not allowed to come out until the sessions are over. Real-life cosmonauts can't just open the door of their spacecraft and go home if they get bored and neither can the Institute's volunteers, or else the experiments would be pointless. The mental strain of being cooped up for so long can make even the professionals go crazy. Russian cosmonaut Valery Ryumin warns, 'All the conditions necessary for murder are met if you shut two people in a cabin and leave them together for two months.' In 1999 an international team entered the IMBP's isolation chamber for a 110-day test. Judith Lapierre from the Canadian Space

Agency found herself being kissed by the Russian commander, even though she had told him to back off. There was also a fight between two Russians that left blood on the walls.

So far no one has used any kind of weapon in space or while training, but it could happen. The Russian Soyuz capsule does contain a couple of lethal items: a three-barrelled TP-82 pistol and a machete knife that doubles as a handle for the pistol. Two barrels on the TP-82 shoot flares so that rescue teams can locate the re-entry capsule if it comes down in rough terrain, while the third barrel fires real bullets. These are only supposed to be used if the crew has to protect itself from wild animals, should their return capsule come to ground in dangerous territory. It wasn't entirely unknown for early Soviet space crews to be harassed by wolves while awaiting rescue from a miscalculated landing in a remote Siberian wilderness. It's possible that someone could start a fight with all this gear. And probably they could end one too. So it's probably best not to have any guns on your ship.

Of course, American astronauts can be just as fragile under strain. In 1973 the third crew aboard NASA's first space station, Skylab, went on strike, because they thought that Mission Control was making them work too hard. Tensions have sometimes arisen on modern space shuttle flights too. 'Imagine taking a trip cross-country with your family,' says the NASA psychologist Mark Shepanek. 'Now imagine that it lasts for several weeks on end. You can't open the windows. You can't even get out of the car. You have to go to the bathroom and eat all your meals in the car. After a while you'll have problems getting along with your family, even though you love them.'

melancholy moon men

Astronauts sometimes experience nasty bouts of depression after coming back to earth. Buzz Aldrin, co-pilot of Apollo 11, had a mental breakdown a year after his historic mission. 'After I had landed on the moon, what else was there for me to do with my life?' he said. 'I felt completely empty.' Other Apollo astronauts had similar problems, as did a number of Russian cosmonauts. The world wondered how crewmen who had been so disciplined up in space could become so unstable afterwards. The word 'astronaut' still conjures up the image of a cool professional with steely eyes, a fixed smile and an unwavering dedication to robotic efficiency. How strange, then, to find that astronauts have just the same failings and passions as the rest of us. They can even fall in love with the wrong people . . .

Lisa Nowak earned her spurs flying F-18 fighters and other high-performance jets for the US Naval Air Systems Command. She capped a glittering career in July 2006 by flying a two-week mission on board NASA's Space Shuttle *Discovery* and docking with the International Space Station. Buzz Aldrin made supportive comments about Nowak, speaking from experience: 'Astronauts are not super-human. They lead ordinary lives and have varied personalities.'

Clearly Nowak had bumped up against what Aldrin once described as 'the melancholy of all things done'. After flying as high as any aviator realistically can in this generation, she came back to earth and discovered there was nothing much left to do that she hadn't already accomplished. With more than a hundred NASA astronauts competing for future space shuttle flights, it became clear that Novak's chances of revisiting space were slender, given the very few opportunities remaining before the shuttle programme ends in 2010. So she diverted her energies into falling in love with a fellow shuttle astronaut and got herself into a surreal muddle, all court cases and cops and press cameras, plus the humiliating end to her career as an astronaut. NASA had to confront a hard truth: space is easier to understand than the hidden landscapes of the human psyche. The same will be true for you. While you focus on the hardware details of your rocket ferry and space hotel, keep a wary eye on the unexpected human element.

disunited states in space

International crews sometimes find that their different backgrounds can cause tension. In 1995 Norman Thagard was the first NASA astronaut to spend time aboard the Russian space station Mir. He found the language barrier hard to deal with. Communications with the ground were unreliable, and he had to wait two or three days at a time before hearing a familiar American voice. Thagard had spent many weeks in Star City, the cosmonaut training complex near Moscow, learning all of the Russian control panels and hardware, but not the people. Once in orbit, the loneliness took him by surprise. His Russian-language skills were OK for technical chatter, but not so good that he could get friendly with his crew mates.

With these cultural problems in mind, McDonnell Douglas, the company responsible for equipping the International Space Station's living quarters, hired a social scientist to study how people sometimes fail to get along in space. Dr Mary Lozano interviewed astronauts from America, Europe, Japan and Russia. Apparently, Japanese astronauts think that Americans often ignore instructions from Mission Control, while Americans think that the Japanese are too cautious when making decisions. Lozano also found that the Dutch and the French enjoy good food and treat mealtimes as a social break. American fliers seem to eat in more of a hurry.

It aslso seemed to Dr Lozano that Russians think Americans are show-offs. In July 1995 the Space Shuttle *Atlantis* brought Norman Thagard home from his three-month stay aboard the Mir station, along with two cosmonauts, Vladimir Dezhurov and Gennady Strekalov. Russian doctors insist that cosmonauts shouldn't try to walk as soon as they've landed. Instead they flop into couches and take it easy, readjusting to earth's gravity after their long stays in weightlessness. But Thagard wanted to walk as soon as *Atlantis* had rolled to a halt on the runway. He didn't mean to upset anyone, but his Mir companions thought he was insulting them.

When it comes to designing your spacecraft these human factors should be uppermost in your mind. All machinery is just a reflection of the cultures that create it. If those cultures clash, then the machines might not work. Control panels for international spacecraft are tricky to lay out. Americans often use the word 'control' when talking about individual switches and levers that make a spacecraft do something, while Russians think of the *people* who control their actions by giving them orders. NASA's famous phrase 'Mission Control' just adds to the confusion. Worse still, the NASA custom of flipping a switch 'up' for 'on' is not always familiar to astronauts from other nations. In Britain it's the other way around. As for all those warning lights, flashing red displays tell Americans that something is wrong, but for Chinese crew members, red can symbolise good fortune rather than bad. Cultural differences can be a problem if astronauts need to make decisions in a hurry.

the lavatory thing

One of the trickiest aspects of life in space is the 'waste-management compartment', the lavatory. A male astronaut can take a leak by fitting a condom-style catheter over his penis, attached to a suction tube. For females, the snug fit of their urine-disposal hose is more of a problem. You can spend ten minutes in the cramped waste compartment doing what you have to do. And then, if things go wrong, maybe another hour cleaning up. So, do try to design your lavatory *really* well, because it's the one thing apart from the nausea problem that's guaranteed to ruin your passengers' experience.

For instance, in 1973 America's first space station, Skylab, had a bathroom that was extremely unpleasant to use. The metal floor was designed as a hygienic wipe-clean surface, so it had no firm footholds, and at washtimes the astronauts slithered about in there like skaters on ice. Having a good wash took hours rather than minutes, because every last droplet of water that strayed outside the plastic shower enclosure had to be mopped up to prevent moisture seeping into electrical equipment and short-circuiting the works. This unexpected extra housework cut into the few precious moments set aside for relaxation. When Mission Control tried to discuss this problem, the frazzled men aboard Skylab lost their temper. The final insult was an unbreakable mirror made from a sheet of polished metal rather than glass. It wasn't up to the job, and the astronauts couldn't see their reflections to wet-shave. So the third and last crew to visit the station stopped shaving. After a few weeks in orbit they looked like a trio of abominable snowmen, crumpled, heavily bearded and, frankly, more than a little crazy.

A nice bathroom is *really* important, especially when you bear in mind the effect of weightlessness on the stomach. During a long mission you may have to take medication to keep your digestion working properly. Sometimes this loosens your bowels too much for comfort. In the waste compartment, you will strap yourself down to the lavatory seat. Instead of water (which would be disastrous in zero-gravity) powerful suction pumps pull the waste into a holding tank. But if the seal between the buttocks and the seat isn't nice and snug, leakages can happen. Then, gobs of urine and solid waste will escape and float around the cabin like swarming insects.

After more than 30 years in the space business neither Russia nor America has yet solved the ultimate cosmic mystery: how to make a space toilet that actually works. A manned spaceship can't afford to store the entire mass of waste produced by its crew. Vents on the side of the vehicle sprinkle the waste into space, after special shredder and vaporiser systems have rendered it into particles as fine as dust. Waste disposal is no joke. One of the weirdest hazards of being in space is the possibility that frozen particles of urine (or worse) dumped from a previous mission might smash into you at 17,500 mph, killing you instantly. The vented waste stays in orbit for weeks or months before atmospheric drag pulls it earthwards. A speck of waste with just 1/36th the mass of an aspirin tablet carries the orbital energy and destructive potential of a 0.3-calibre bullet. In ten years of operation, Russia's old Mir space station lost 40 per cent of capacity from its giant solar cell arrays due to tiny orbital impacts. Samples of the cells retrieved for analysis showed a high proportion of the dents came from flecks of human waste.

Waste systems often break down. On one occasion a NASA Space Shuttle commanded by Bob Crippen developed an 18-inch long 25 lb solid crystal of frozen waste fluids, hanging off the vent and clogging the works. Crippen warned that he might have to bring the shuttle home early, since the crew cabin was becoming distinctly unhygienic. It doesn't help that there's no room for a shower on a shuttle. The best anyone can do is to rub themselves down with foil-sealed moistened towels, like airline handywipes. Just what you need after a hot and sweaty five-hour spacewalk. There's nothing like a hot bath to help you freshen up – and on today's available spacecraft and space stations, there is, indeed, nothing like a hot bath.

death rays

It's often said that the greatest hazard for any space voyager is the radiation. Well, yes – and also, perhaps no. Cosmic radiation – the mysterious sub-atomic particles produced by exploding stars, which hurtle across the universe at close to the speed of light – can destroy biological cells. Among the scariest particles are neutrinos. They have almost no mass, but they carry colossal energy. They are so small and fast that 99.999 per cent of all neutrinos that hit the earth pass straight through. Literally, they slip through the gaps in the atoms that make up our world. The only way of detecting neutrinos is to catch the occasional straggler, using detectors buried at the bottom of mineshafts. Unfortunately, if a neutrino *does* interact with something, its energy causes enormous disruption. Astronauts have occasionally reported sudden flashes of light in the backs of their eyes, so brief they hardly thought them important. Scientists now realise that these are neutrino impacts. Under an electron micro-scope the helmet visors of some astronauts have revealed laser-like paths where the neutrinos have ploughed through the molecules of the perspex and into the astronauts eyeballs, then through their skulls and out through the backs of their helmets. If the trail of atomic destruction in the perspex is anything to go by, the neutrinos will presumably have wreaked irreversible damage in the much more delicate fabric of the astronauts' brains. Unlike the rest of the body, brain tissue does not repair itself.

Nevertheless, radiation is surprisingly not as much of a worry as you might think. It's an added long-term risk which your clients might well decide is worth taking, provided you give them the right information. Airliner pilots are routinely exposed to a good deal more radiation than space travellers. It's not just the strength of the radiation, but how often you're exposed to it that counts. Space travel works out safer, because that 'once in a lifetime' trip is just that: a risk you'll take once (or at a pinch, two or three times) in a lifetime. High-altitude airliner pilots soak up radiation every day of their working lives over perhaps two or three decades, so they're more at risk of damaging their DNA by the time they get old. Suborbital space flight exposes you to minute amounts of additional radiation, typically enough to cause only 0.01 fatal cancers per 10,000 people. By comparison, 50-year-old white males suffer cancer at the rate of 700 per 10,000 people, with 200 of them fatal; among 30-year-old women of all races, the cancer rate is 150 per 10,000, with 20 fatalities.

You can assure your clients that deadly space radiation is not a concern for short-duration space adventures. When it comes to longer missions, for instance voyages to the moon or Mars, then the risks increase, but again all that happens is that a space explorer's chances of developing cancers in later life shift upwards from the average. Leaving aside those unfortunate few who have been prematurely snatched from life by accidents, none of the people who have flown into space so far have died under circumstances that suggest they would *definitely* have lived longer if only they hadn't chosen 'astronaut' as their job description. The added risk is hard to separate from the overall statistical incidences of cancer in the world at large.

what that floating feeling does to you

There is one thing that space explorers from all nations can agree on: living in orbit for long periods isn't especially good for your health. The most serious medical problems caused by weightlessness happen deep inside the skeleton. On earth the constant shock of walking, running or just standing still seems to stimulate your bones into replenishing themselves. In space, those ol' bones shed calcium at an alarming rate and you are afflicted by something similar to osteoporosis, a condition usually associated with growing old. Essentially, your bones become brittle in space. The shed calcium circulates in your bloodstream and a heavy proportion accumulates elsewhere, for instance as kidney stones.

As well as bones, your muscles will also atrophy, partly through complex biochemical side-effects, but also through sheer lack of use, especially in the lower body. You won't be using your legs and feet much in orbit. Professional astronauts call this the 'chicken-leg syndrome'. In contrast, your arm muscles will retain their vigour, because you'll be using them for pulling and pushing yourself through hatchways and along access corridors. And your cardiovascular system also changes. Blood pools in the head, causing a puffiness around the eyes, headaches, drowsiness and nasal congestion. Your body's internal regulation systems will misinterpret all this as an excess fluid build-up and you might have the sensation that you need to pee. Thirst levels are dampened and inadvertent dehydration becomes a danger unless someone reminds you to drink regularly and often. Blood cell production within the bone marrow is reduced, leading to anaemia and a weakening of the immune system. The heart atrophies and becomes used to a much-reduced pumping load, which can be very dangerous when your return to earth pushes the load up again.

Your sense of balance also goes haywire. Intense nausea is common among even the hardiest space crews, at least in the first few hours of weightlessness. Incompatibilities between the inner ear's balance and the visual senses can cause strange illusions. But there are some indications that the body eventually adapts, finding a new balance in the unusual environment of space (along with the senses). Crews that have been in space for just a week or so are often much weaker on their return than those that have been up for many weeks or even months. What's more, astronauts have long made use of certain exercise regimes to simulate the bone shock required to keep the skeleton healthy and to give their hearts plenty of work. Certainly, Russian cosmonauts that have flown for a year or more find returning to earth pretty uncomfortable, but they recover their usual state of health in a matter of weeks, as far as anyone knows. What are the long-term effects of spaceflight? These have yet to be determined, but there is evidence that a slight overall reduction in health can result.

So, why not become a medical pioneer? The bone disorders that afflict astronauts in weightlessness may well justify the expense of sending them into space in the first place. Nearly 30 million Americans suffer from osteoporosis. The total costs of medical treatment associated with this condition amount to perhaps $100 billion a year: as much as the entire cost of the International Space Station's development. If you or your weightless colleagues can be studied, revealing more secrets about our bone chemistry, then your space project may eventually pay for itself by contributing significantly to medical knowledge.

Osteoporosis usually develops over some years, but in space similar symptoms appear to be greatly accelerated. As early as the second day in orbit an astronaut's bloodstream will show high levels of calcium shed from the bones. But the skeleton is a dynamic structure, constantly replenishing itself. New calcium from food intake is collected and harnessed in the bone matrix. Osteoporosis happens when the replenishment of calcium tries and for some reason fails to keep pace with the rate of loss. On earth the long timescales involved in these processes make rapid medical discoveries difficult. Up in orbit your human lab rats can vary their exercise regime, diet or any number of other factors and rapid short-term changes in their bone processes are measurable over days and weeks rather than years. In fact, all of the side-effects induced by weightlessness could be exploited in order to advance our understanding of the human body. If you can come up with tangible data, your space station might actually make you rich.

six the moon is a harsh mistress

In this chapter we'll look at some of the grander space projects on the horizon, mega-projects that will require many billions of dollars of government funding before they can get off the ground. We're talking now about missions to Mars and even to the asteroid belt between Mars and Jupiter. At first glance you might think that there's no way these mega-projects can have any relevance to your modest company, but you'd be wrong. You probably won't be building the principal ships for these missions and also you probably won't be lining up any of your passengers for pleasure jaunts to the far planets (the trips we are about to discuss are for hardened professional astronauts only), yet there is still a good chance that your little ships can reach a nearer target: the moon. Failing that, the government space agencies are on the lookout for a wide range of support activities on the ground, and in space.

Starting at the most basic level, does your three-people-and-a-pet-dog company have, perhaps, a new and slick management software tool that could help a space agency plan the logistics of a huge project? Subcontracting this kind of work to private enterprise is the lifeblood of a space agency. In fact it is obligatory to put these tasks out to tender, because one of the responsibilities of a tax-funded agency is to spread the benefits around. You don't have to be a huge company to pick up some of that work, especially when you're offering brainpower rather than costly industrial manufacturing expertise. Moving upwards a notch, NASA has initiated a wide range of prize funds designed to stimulate new thinking about how to do space flight. We'll examine these in a bit more detail later. Finally, if you are a somewhat larger-scale operator, there is a need for real space hardware at the less obvious end of the dramatic curve: not the ships themselves, but all the thousands of bits and pieces inside them that make them work. And let's not forget the catering side of things. After all, even hardened professional astronauts have to eat. Anyway, let's look at the stories behind these mega-projects.

to the moon – again

Disasters set us all back, but sometimes they force us to reinvent ourselves and come up with a new and better plan. Certainly this is true of NASA. First the space shuttle *Challenger* blew up almost immediately after lift-off in January 1986, then the space shuttle *Columbia* broke apart in the atmosphere in February 2003. To lose one shuttle and its crew might seem unfortunate, but losing two was a real wake-up call. The space shuttle has certainly achieved great things and it is a true marvel of engineering. Yet when NASA and the world's press celebrated it, in the early days, as 'the most complex machine in history', so many people failed to see that its complexity was neither impressive nor wonderful. It was a dangerous flaw. The shuttle has been a science-fiction dream come true, a marvellous winged spaceship flying time and again into orbit and coming home to land like a plane. But sometimes, like so many such dreams, it has delivered nightmares too.

On 14 January 2004 President George W. Bush made a televised speech during a special visit to NASA's headquarters in Washington DC: 'Today I announce a new plan to explore space and extend a human presence across our solar system.' The first goal, he said, was to improve the surviving shuttles (there are only three left out of an original five-strong fleet), get them flying again and finish America's pivotal contribution to the International Space Station. No surprises there. But as his speech continued, radical new ideas emerged. 'Our second goal is to develop and test a new spacecraft, the Crew

Exploration Vehicle, by 2008, and to conduct the first manned mission no later than 2014. Our third goal is to return to the moon. Using the Crew Exploration Vehicle, we will undertake extended human missions to the moon as early as 2015, with the goal of living and working there. With the experience gained on the moon, we will then be ready to take the next steps of space exploration: human missions to Mars and to worlds beyond.'

These plans emerged after months of secret consultation between White House officials and space experts during 2003. Most politicians agreed that NASA needed a new sense of purpose and Mars was an obvious long-term goal for human space flight, yet Bush had personal reasons for being cautious. Back in 1989 his father George Bush Snr celebrated Apollo 11's twentieth anniversary with Neil Armstrong, Buzz Aldrin and Mike Collins standing at his side. 'For the 1990s, we have the Space Station,' he said. 'For the new century ahead, we should go back to the moon.' Then, in words that were music to NASA's ears, he talked of 'a journey into tomorrow, a journey to another planet, a manned mission to Mars'.

Short of setting a precise target date, he seemed to have given them the green light. A huge NASA team set to work on an infamous document nicknamed the '90-Day Report', so called because this was how long it took to deliver a plan almost guaranteed to make Bush Snr wish he had never mentioned Mars. An expanded version of the Space Station was to service a thousand-ton interplanetary craft assembled in earth orbit. The Mars return trip would take 18 months, with only a few weeks spent on the surface: just enough time to plant a flag and snap some photos before heading home. NASA priced the project at just under $200 billion over two decades, but many financial analysts expected the figure to be more in the region of $500 billion. This budget appalled everyone and the whole idea was scrapped. Flash forward again to 2004. The son didn't want to repeat his father's mistake, and Bush Jnr made only vague references to a Mars mission in his 14 January speech. But when it came to the moon, he was more confident about possible dates. After all, NASA already *knows* how to get there. It should be affordable and our next-door neighbour in space is less than three days' flight time away. And talking of timing, Bush's speech was influenced by the presidential elections coming up that November. The 9/11 catastrophe and the Iraq crisis defined the first term of his presidency to the exclusion of almost everything else. Like President Kennedy before him (faced with a disastrous failed invasion of Cuba) Bush would rather be celebrated in the history books for some-thing other than bad news; and just as JFK found refuge in the Apollo lunar-landing project, so Bush also turned to the moon as a diversion from his earthly woes. 'We choose to explore space,' he said, 'because doing so improves our lives and lifts our national spirit.'

rearranging the pieces

If Bush's vision is to become a reality drastic changes will have to be made within NASA. Construction of the Space Station will end in 2010, so that funds can be diverted towards the new lunar project. Russia, Europe and Japan will then be responsible for the Space Station. At the same time, the space shuttle is to be retired for ever. After 2010 the Crew Exploration Vehicle (CEV) highlighted in Bush's speech will become America's celestial flagship. But this time NASA has to listen to outside industry, including modest-sized space operations such as yours. The CEV project will need a lot of smaller-scale support.

Be prepared for this unexpected new system. The CEV is a reversal of everything NASA has worked towards in the last 25 years. The dream of an all-purpose winged shuttle, flying cheaply and regularly like a commercial cargo plane, never came to fruition. Whenever a space shuttle puts a satellite or a space-station module into orbit the costs are huge, because astronauts always come along for the ride. The launch weight given up to life support and crew cabins has to be subtracted from the cargo, and that limits how much payload a shuttle can carry. It's more efficient, say the CEV's designers, to haul cargo in disposable uncrewed rockets and launch astronauts separately in a small module that carries people and nothing else. There's no need for the space shuttle's huge wings and cargo-bay doors in the people-carrier bit. All that's left to worry about is the crew cabin at the front. Instead of the space shuttle's complicated aeroplane-style cockpit there will be a simpler, self-contained and cone-shaped crew capsule that fits on the end of a solid-fuelled rocket, closely based on the existing side-slung solid boosters that help propel the space shuttle into orbit.

The CEV capsule also serves as a neat escape pod if anything goes wrong during lift-off. Basically, it pops off the top of a malfunctioning rocket like a cork and parachutes back down to safety a few thousand metres away from any launch-pad inferno. Similarly, if things foul up at high altitude the capsule can blast clear of a wayward booster. In short, the CEV is an adaptable component in a building-block system. It plugs onto booster rockets, fits on top of lunar-transfer modules and landers, and it can even dock with the Space Station – or your own privately designed modules and support vehicles. But is this progress or a weird leap back in time? Robert Zubrin has worked tirelessly over the last decade to promote a human Mars mission, and he's not sure. The topsy-turvy logic of space planners often exasperates him: 'In the 1970s, when President Nixon killed the Apollo programme and ended lunar exploration, NASA said we would do all that again one day, after we had developed cheaper transportation to orbit using a winged shuttle. We've spent 25 years trying that, with no positive results. The shuttle costs more to fly than the Saturn boosters we had for the Apollo missions. Now the shuttle is going to be replaced with a CEV capsule that looks and flies just like the old Apollo!'

learning from the past

Apollo may be old, but it still sets an example that is pretty hard for modern designers to match. In 1962 NASA engaged in a passionate internal argument about the best way to reach the moon with its Saturn V booster or some variation of it. The long-held dream of simply blasting a rocket to the moon and bringing it back home again was a staple of science fiction and it couldn't help but infect the imaginations of real-life rocketeers. However, the technical problems of this approach would have been stupendous. The upper stage of the lunar rocket, about the weight and length of a Navy destroyer, would have to land on the moon stern-first without toppling over, then take off again without the benefit of a launch gantry and ground crew. It would have to carry all the fuel and equipment for the return voyage, including heat shields for the punishing 25,000 mph re-entry into the earth's atmosphere. Since all this gear went down to the lunar surface and then came all the way up again, the fuel and weight requirements for the whole thing were simply staggering.

An alternative plan was hatched and NASA devised its famous lunar-landing module, the first human-carrying vehicle designed to operate purely in the vacuum of space. Unlike Apollo's main conical crew capsule, it had no heat shielding and no need for an aerodynamic shape. Most of the outer skin was nothing more than lightweight metallic foil, and even the small two-man crew compartment could easily have been punctured with a screwdriver. The ship was so delicate that the fuel and oxygen loaded aboard prior to a mission accounted for more than 70 per cent of its weight.

The bug-like craft was built in two halves. On completion of lunar-surface operations, the landing legs, main rocket engine and empty fuel tanks in the lower section were abandoned to save weight. Only the compact crew module lifted off for the return trip, using a small reserve fuel tank and engine. If the ascent engine had failed, the astronauts would have been stranded on the moon for ever. For safety, the engine had only one moving part, a valve to let in the fuel. For extra reliability, it used a hypergolic fuel and oxidiser combination that burned immediately when mixed, without requiring an electrical spark for ignition. After making a rendezvous with the command module in lunar orbit (no easy task) and transferring its crew, the ascent stage was thrown away altogether. It drifted in space for a while and eventually plummeted down to the moon and smashed to pieces, alone and forgotten, among the silent hills and craters.

One day the abandoned but still intact lower stages will be visited by tourists. Nearby, they will find the Apollo astronauts' original footprints, looking almost as fresh as the day they were imprinted in the lunar dust. There is no wind on the airless moon, no rain, and almost no erosion to sweep away the traces. It is possible that the footprints will survive for thousands of years, although solar radiation, micrometeorite impacts and the gradual stresses of heat and cold in that harsh environment will eventually wear them away. The lunar module remnants should last a million years or more, provided souvenir hunters don't pick away at them like vultures. Further afield, more adventurous explorers will be able to follow the wheel marks made by battery-powered rover vehicles, carried on the side of the lunar module descent stages during the last four Apollo missions and used to transport astronauts across the surface. These rovers, also perfectly preserved, are parked alongside the descent stages, but their tracks, criss-crossing back and forth from the touchdown sites, stretch over thousands of yards of terrain.

The upcoming missions will follow a remarkably similar profile, albeit with somewhat larger and more rugged ships. Each CEV will dock in earth orbit with an unmanned transfer vehicle, launched separately and carrying a spindly landing craft. The combined CEV and lander will then blast out of earth orbit and head to the moon. Once in lunar orbit the lander will then detach and drop down to the surface. At the end of a mission the upper stage of the lander will be boosted back into space by a small ascent motor. It docks once again with the CEV mother ship and the crew transfers to the capsule for the return trip to earth. It all sounds spookily familiar from the old days of Neil Armstrong, Buzz Aldrin and their Eagle lunar module. No wonder some NASA veterans call the CEV 'Apollo on steroids'. One version of the plan calls for ferry vehicles to switch constantly between earth and lunar orbit, carrying CEVs and lander modules back and forth. Similar hardware will deposit moon-base compo-nents ahead of the astronauts' arrival. Inflatable TransHab-style living quarters can be adapted for the moon. It won't be difficult to set up a base. NASA has been studying how to do this for many years.

Private involvement will be crucial. The influential Virginia-based company, Transformational Space, made its pitch to NASA almost immediately after Bush's announcement, urging NASA to 'Base orbital and lunar infrastructure on commercial firms selling services to NASA, rather than the federal government designing and actually owning moon habitats, mining facilities, greenhouses and power stations. In other words, aim for an American-style open frontier rather than a Soviet-style government compound.' These words may actually have had an impact. The CEV and its boosters will probably emerge from the giant Lockheed and Boeing factories, but a great deal of support hardware will come from smaller companies. As for the actual lunar surface operations: at the moment there's still everything to play for. Bigelow, for instance, would obviously be in a strong position to provide those inflatable habitats. Your company might want to look at the wheeled rovers or the autonomous robotic explorers that will support the manned missions.

But is the moon the right place to go? Do we need a human presence there? According to Bush and his supporters, we must gain experience of living on the moon before we even consider doing the same on Mars. And anyway, the moon is a worthwhile target in itself. Bush's speech in 2004 wasn't exactly poetic, but he was right when he said, 'In the past 30 years, no human being has set foot on another world or ventured farther into space than 386 miles, roughly the distance from Washington DC to Boston, Massachusetts. America has not developed a new vehicle to advance human exploration in space in nearly a quarter of a century. It is time for America to take the next steps.' The hardware for that next step will be designed soon enough. The bigger challenge will be to find the money.

After 2010 the CEV project kicks into high gear and serious mission hardware will start to emerge. NASA's annual budget is $16 billion. Its work on the Space Station project absorbs $2 billion of that, and shuttle flights a further $6 billion. Cancelling NASA's burdens on those projects after 2010 will release $8 billion a year. Spending on the CEV between 2010 and 2020 could amount to around $86 billion: roughly equivalent to the cost of the Space Station over the last decade. Things are definitely moving moonwards, and NASA is eager to hear from private entrepreneurs such as yourself. There are opportunities to develop lunar bases, wheeled rovers and long-duration spacesuits, then lease them to NASA, while at the same time exploiting the hardware yourself for lunar tourism, commercial-scientific partnerships and, of course, prospecting for rich mineral pickings.

digging for lunar gold

Supporters of President Bush's space plans claim that the moon's natural resources can be profitably exploited. Some studies suggest that lunar helium-3 (He-3) – deposited in the topsoil by charged particles in the solar wind – could be worth the effort of refining it.

Just 20 tons could power all of North America for a year. The trouble is, fusion reactors haven't come on-line yet and no one is certain that helium-3 could be collected without shifting thousands of tons of lunar soil. You'll be taking a big risk, but the prize for victory could make you the Bill Gates of global energy. Fusion-reactor prototypes on earth are getting (a little) better with every passing year . . .

In the early days, a moon base may be better just for pure research rather than commerce. Since the dawn of the Space Age half a century ago, many astronomers have dreamed of building telescopes and radio dishes inside deep craters where they are shielded from solar light pollution or unwanted 'noise' from earth's countless radio transmitters. If you can do this, you should find hundreds of academic institutions on earth willing to pay for access to your telescope, just as they now sign up for precious time slots with the orbiting Hubble Space Telescope. But be warned: not all astronomers see the benefits. The moon may well be fascinating for mineralogists, but as a telescope platform it offers mainly dirt and gravity, both of which degrade performance in comparison to free-flying space instruments like the Hubble. Then again, no one's tried anything similar on the moon yet. So why not be the first? Your best bet is to land a robotic telescope as an experiment and see how well it performs. If all goes well, you will have made a good case for building a larger 'scope staffed by eager astronomers.

In which case, you'll need a plot of lunar land. Before committing yourself, watch out for the legalese. At the moment no one can claim any stake up there. Just like the Antarctic, the moon is regarded as a territory that cannot be 'owned' by anyone. This could make life difficult if you are hoping to exploit mineral rights up there or build a permanent settlement on a profit-making basis. Quite literally, you might not have the right to do anything of the sort. The 1967 Outer Space Treaty says that no one nation can claim ownership of any celestial body, including the moon. However, the laws are vague when it comes to private rather than national expeditions, and even vaguer when it comes to multinational private companies . . . In 1984 a second treaty agreement (known as the Moon Treaty) was drawn up specifically to ban private ownership of lunar real estate. Some countries signed up to this, while others did not. No one is absolutely sure what the legal status of the moon really is. Those websites offering to sell you an acre of lunar landscape are just kiddie toys. They have no legal standing whatsoever. If you're serious about setting up a business venture on the moon, then consult the space lawyers.

where to place your base

Peary Crater near the lunar North Pole is named after the US explorer Robert Peary, who reached the Arctic North Pole for the first time in 1909 after a trek of 37 days. Some critics have since suggested that he may have ended his journey at the wrong place, so he doesn't deserve the glory of having a lunar crater named after him. All you need to know is that Peary crater is the best bet for your first moon base. Several areas on the rim of Peary are almost continuously illuminated by sunlight. From the point of view of settlers living here, the sun will always be low in the sky. Solar radiation strikes these regions obliquely, keeping temperatures at a steady minus 58° Fahrenheit (–50° Celsius). Antarctic research stations routinely operate in similar conditions.

Just as significant are the nearby deep-walled craters where sunlight *never* penetrates. It's possible that fragile water ice and other remnants of ancient cometary impacts still survive in the deepest, coldest shadows. You can extract drinking water and oxygen, plus hydrogen to power your ferry rockets for take-off. There are many craters where pockets of perpetual darkness may exist, but so far, none are bathed in constant daylight apart from Peary crater. So this site may offer you the best of both worlds. It's permanently light, yet within easy reach of perpetually shadowed sites where valuable ice may lurk.

Permanent living quarters can eventually be constructed under protective blankets of lunar topsoil. This is a cheap and simple way of shielding against micrometeorite impacts, while keeping control over temperatures inside your living quarters. But you won't want to just sit inside your lunar home for months on end. Long-range wheeled rovers will be essential. Docking hatches will allow them to link up and create temporary exploration camps far afield. The rovers will be fully inhabitable extensions of your main base. They will have exterior robotic grappling tools so that your scientists can study rocks and soil samples without always having to climb into their spacesuits.

That's the plan. The first thing you might try is to beat NASA to the moon! OK, not all the way down to the surface, but . . .

voyage round the moon

The USSR's trusty Soyuz capsule was designed originally to be part of a 1960s lunar-landing project. Technical and political confusions and disastrous problems with booster rockets prevented the Russians from matching the achievement of Apollo 11. The Soviet appa-ratchiks decided after a while to pretend that the mere idea of trying to land men on the moon had never even entered their minds and that earth-orbiting space stations had always been their priority: a much better way of displaying Soviet space prowess. In recent years the extent of that lie has been exposed. A gigantic effort to reach the moon really was undertaken. Three or four of them, in fact. The major one was supposed to use a rocket called the N-1, every bit as large as America's Saturn V booster. Unfortunately this giant was beyond USSR's ability to build successfully. (Remember KORD trying to control all of those engines?) The N-1's unmanned test flights were explosive failures.

Meanwhile a rival project used a slightly smaller hypergolic-fuelled rocket called the Proton. There was a plan to send a Soyuz capsule with just one crewman around the moon, without attempting an actual landing. If the Soviets had managed to pull this one off ahead of NASA's Apollo, it would have been yet another coup for the Communists. Accordingly, an unmanned Soyuz derivative (code named 'Zond') was sent around the moon, and its capsule returned safely to earth. Thoroughly alarmed, NASA hurriedly sent its Apollo 8 crew capsule on a similar trajectory in December 1968, around the moon and then back to earth without actually touching the surface. The lunar-landing module was left back on earth because it was still several months away from readiness, but NASA was scared by the Zond tests and wanted to make absolutely sure that the USSR couldn't sneak in a last-minute PR victory. That was more or less the point at which the Russians decided that the moon race was over.

So how about reviving that ancient Soviet dream and sending a Soyuz around the moon? The Virginia-based company Space Adventures, responsible for sending the first private space travellers to the International Space Station, is looking very seriously at this idea; likewise, a company called Constellation Services International. American cash allied to Russian hardware could make this circumlunar mission possible in a very few years' time.

The Soyuz has an egg-shaped docking module attached to its nose. The lunar plan calls for one more small cylinder to be clipped onto the front of the egg, providing just a little bit more fuel, water, food and living space for the six-day lunar round trip. Finally, a 'kicker' rocket gets the combined craft out of earth's orbit and pushes it towards the moon. Your stretch-limo version of Soyuz then flies what's known as a 'free return trajectory'. Just for once the technical term is a perfectly lucid description of what it means. The return to earth is a free ride, requiring no rocket ignitions. Obviously this is great from the safety point of view, because you don't have to worry about engine failures stranding your hapless passengers in eternal orbit around the moon. It's also great that you don't have to carry any fuel for the return trip.

How does this magic trick work? Once you have escaped earth orbit at the necessary speed of 25,000 mph, Isaac Newton takes over the driving. As the spaceship nears its target, lunar gravity gets ever more of a grip on it and whips it around the far side of the moon. If you applied a braking burn at this point, firing your rocket against your direction of travel, you could slow down and go into lunar orbit. But if you just leave things be, then your ship is hurled about three quarters of the way round the moon before escaping once more into space. It's not a complicated matter to ensure that this escape sends the craft back towards the earth.

The hard part comes at the end, during the crew capsule's re-entry into the atmosphere. Ordinarily, dropping down from earth orbit, you will hit the upper reaches of the atmosphere at around 17,500 mph. But coming back from the moon your impact speed will be more like 25,000 mph. The 1960s Apollo crew capsules were designed to handle this, but Soyuz is not. You'll need an additional 300 lb of heat shielding, plus a 'double-bounce' re-entry path to shed some of your speed before final descent into the atmosphere.

Essentially, you skip the capsule over the top of the atmosphere – just as a flat-bottomed stone can be skimmed over a pond very fast – only to slow down and sink after a couple of bounces. Bringing a capsule back from the moon and getting it into exactly the right angle of attack for re-entry has been compared to aiming a rifle bullet to hit the edge of a piece of paper a mile away. If the angle of attack is too steep the capsule will burn up like a fiery meteorite. At too shallow an angle, the first bounce off the atmosphere sends you skittering back into space, and you might become stranded forever in an elliptical (oval-shaped) orbit around the earth. But don't be put off. Re-entry experts are good at planning this sort of manoeuvre. It works almost every time.

landing a prize

At a more modest scale, your company might want to enter a competition to design a miniature unmanned lunar-landing vehicle. NASA and the X Prize Foundation, working in alliance with the Northrop Grumman Corporation (builders of the original Apollo lunar modules in the 1960s) with the Wirefly mobile phone company as additional sponsors, are offering upwards of $2 million to anyone who can make a miniature prototype vehicle hover and land (on earth) under certain constraints. It doesn't matter, for now, if these are just glorified toys. The point is to explore stuff like automated radar touchdowns and to test new kinds of fuel mixture, try different shapes for landing struts, engine nozzles and so on. Although it'll probably be the usual cabal of manufacturing giants who get to build the *real* lunar landers of the future, small companies, and even individuals, really can get in on the act by coming up with useful ideas (and, of course, a small company with a great idea might grow into a big company later on).

The annual Wirefly X Prize Cup competition is held each October at Holloman Air Force Base outside the city of Alamogordo, New Mexico, with the aim of showcasing new private space technologies. Competitions are a big part of the attraction and the Lunar Lander Challenge is the most popular of all. But so far it has been one of the toughest to crack. One story in particular illustrates the fiendish complexity of the control systems that you need to master, even at this junior-league level.

In the summer of 2007 Armadillo Aerospace, a well-respected company from Texas founded by *Doom and Quake* PC games entrepreneur John Carmack, persuaded a little craft called Texel to hover about 20 feet off the ground, then descend gently to a landing. All appeared to go well, except that after Texel touched land again its engine wouldn't shut down. Instead it kicked up another burst of thrust and hurled Texel back into the air. A safety tether dangling from a tall crane was supposed to prevent Texel from skittering off at a tangent and perhaps injuring onlookers. Now it shot upwards so rapidly it threatened to dash itself to pieces against the top of the crane. Armadillo's technicians shut down the wayward engine by remote control and Texel fell like a stone. On impact with the ground, a fuel tank ruptured and a fireball engulfed and destroyed the entire machine.

detective work

If the Armadillo team was to learn anything from this unhappy day, it was only through a careful investigation of Sherlock Holmesian proportions. Tracking through their reams of data they identified the unfortunate chain of events that led to the crash. Texel contained an inertial guidance system linked with a GPS sensor, which kept track of its position over the ground: essential if the craft was to seek and then land on the exact target zone 300 feet away from the original lift-off point, as specified by the Wirefly competition rules. Mechanical sensors on Texel's legs should have caused the engine to shut off as soon as the ship touched the ground, but just as this routine was kicking in something else went wrong.

The gentle shock of the landing jarred the guidance sensors and made them pulse a false signal. Texel's master computer received two signals, one saying it had landed and another saying it was *about* to land in the wrong place. The 'wrong place' signal won through because the 'touchdown' signal was just a few amps weaker than the computer had been programmed to expect, and therefore was treated as unreliable. So the engine was throttled back up to full power, because Texel thought it was still falling through the air and needed that extra spurt of thrust for one last hover. All these errors happened within the span of a few seconds. Imagine how difficult it must have been for Armadillo's horrified staffers to react as this drama unfolded with such alarming speed.

The lesson from this – and from countless similar stories throughout the history of the Space Age – is that spacecraft are complicated beasts; their failures are almost invariably high-speed and tend towards the explosive, and the underlying causes are subtle and hard to track down.

just so you remember

Like so many modern space pioneers you'll probably find yourself trawling through the vast and undeniably inspirational archives from NASA's Apollo programme to find out how they coped with similar bad days. On 4 April 1968 the gigantic Saturn V booster took off on an unmanned test flight. Two of the five F-1 main engines conked out shortly after lift-off, leaving the rocket dangerously unstable. Remember the VTOL problem that we encountered earlier: maintaining a symmetrical pattern of engines if one or more of them fails? This time the 'on' engines were biased too much to one side. Mission controllers kept their fingers over their abort buttons, because at one point (too high in the trajectory for civilian ground observers to notice) the rocket keeled over and headed nose-down towards earth before levelling out again. Somehow its guidance computers compensated for the failed engines and the upper stage, with its unmanned Apollo capsule, just about staggered into orbit.

Thousands of engineers worked on fixing the Saturn. Analysing the failures was tricky, because there was no wreckage from the launch to examine (the expended first stage was beyond salvage at the bottom of the Atlantic). The only evidence was radio telemetry data describing the flight in numbers and symbols on computer print-outs. From this slender evidence an incredible detective effort resolved the problem. Kerosene and oxygen fuel thrumming at high speed through the pipework of the five main engines had set up a resonance that shattered the fuel lines. It's a bit like the effect you get in a tiled bathroom when you whistle at a certain pitch and the whole room suddenly hums. Make the tiles out of several different materials at random, and you can eliminate the humming. With a similar theory in mind, NASA redesigned the F-1 engines using a new mix of materials and the vibrations in the fuel lines were eliminated. They also pumped a small percentage of inert helium into the fuel mix to dampen the frequencies as the fuel swooshed through the pipes. It was a brilliant effort and probably one of the cleverest (yet least-publicised) engineering achievements in history. Ground tests confirmed that the fix had worked, but it was still an incredible risk to launch the next Saturn V with men aboard: and all the way to the moon at that.

So here's some advice: read everything you can about the Apollo project, from its politics to its design and execution. It's lesson after lesson, with some more lessons thrown in for good measure.

how you can touch the moon

In September 2007 Google co-founders Larry Page and Sergey Brin (in cooperation with Peter Diamandis of the X Prize Foundation) announced yet another major space prize: 'Today we are challenging private teams from around the world to design and build robotic explorers and race them to the surface of the moon.' The prize fund is $30 million (actually the $30 million is segmented into a $20 million grand prize, a $5 million second prize and a $5 million bonus) and the deadline is 31 December 2012. This Google-sponsored problem is slightly different to the one addressed by the Grumman/Wirefly Lunar Lander challenge, which tests new techniques (albeit in miniature) that might one day help land astronauts on the moon. Google, by contrast, is demanding an immediate working payload, a robotic rover that takes images and beams back data from the lunar surface at a high bit rate.

Coincidentally, in the same week that Google made its announcement the Japanese Space Agency JAXA launched *Kaguya*, a three-ton $480 million orbiter, the largest and most lavishly equipped object to travel moonwards since the Apollo era. Packed with 14 science experiments and survey cameras, it also carried a pair of microsatellites ready for release into different orbits once *Kaguya* reached its target.

thinking small

And talking of microsatellites, these could be a brilliant way to get in on the act, because you can hitch a lift on someone else's bigger mission. Play your cards right and they'll carry you practically for free. These days it's possible to create satellites and space probes so tiny and light that they can be steered by thrusters smaller than a sugar cube. Essentially this is a 'solid-state' design with no moving parts, built from specialised microchips. The thrusters are incorporated into the circuitry. Tiny heating elements in each thruster warm up a pellet of ammonium salts and vapour from the salts is expelled through a pinhole-sized nozzle. The thrust is only a few fractions of an ounce, but this is sufficient, because the entire spacecraft weighs less than a couple of pounds.

Another thrust option is called an 'ion drive'. Heated elements generate electrically charged particles, and electro-magnets send the particles hurtling away. Again, the thrust is tiny, but it can be sustained for weeks or even months, accelerating the little craft to tremendous speeds. The mini-ship navigates with gyros which use lasers instead of spinning wheels. The entire path of the lasers is deflected and measured within the confines of a specialised chip. As the probe twists and turns the beams take slightly more or slightly less time travelling around a miniaturised set of mirrors and prisms. From these billionth-fraction of a second changes a computer works out how the craft's orientation has altered. You can buy these chips off-the-shelf.

A tremendous amount of mapping, mineral surveying, naviga-tional-beacon support and other tasks can be accomplished from lunar orbit, using microsatellites; and you can easily afford to create some of those missions, especially if someone else is providing the launch vehicle and the transfer stage that gets you from the earth to the moon.

so why send humans?

Many scientists say that human space travel is pointless. Robot probes are cheaper and better, we are told. No one cares if a bleep-ing hunk of metal gets lost in space. Why send astronauts on dangerous and costly missions to the airless moon or inhospitable and distant Mars? Why waste all that money on giant spaceships when small, disposable machines (including microsatellites) can do the job just as well or even better – and at a fraction of the cost?

Certainly robots can be impressive. NASA's wheeled rovers, *Spirit* and *Opportunity*, roved Mars for more than three years and all for little more than the cost of a Hollywood blockbuster. The problem is that these clever toys only covered a few yards of territory a day. Imagine what a human could achieve just by strolling around on Mars for a few days. Robots are not intrigued by a strange glint on a rock. They will never, on a sudden whim, turn over a boulder and look underneath to see what might be there. They can only do what they are told by their controllers back on earth. We need to send astronauts to explore the moon and Mars properly, because only humans know how to be curious. That's the fashionable pitch these days, at any rate. Given the growing emphasis on human missions, your company might want to help those astronauts sort out their wardrobe.

what not to wear

A spacesuit is like a miniature spaceship, complete with everything that an astronaut needs for survival: oxygen to breathe, water to drink and a system for controlling the interior temperature. A backpack holds the oxygen and water, along with a radio, a battery and a water-cooling pump. A smaller compartment on the top of the backpack (usually known as an 'oxygen purge system') holds an emergency reserve of breathing oxygen, which can be activated instantly if the suit is accidentally punctured. This supply hisses through the suit at several times the normal rate, keeping pace with any air escaping from the puncture. Obviously, the purge reserve runs out quite fast if activated, but it makes you feel better having one strapped to your back. Also, inside the suit, there's a garment similar to long johns, threaded with thin, hollow tubing. Water from the backpack cooling system circulates through the tubes, absorbing excess heat. Then the water is pumped round again so that the heat can be lost into space from a small radiator on the backpack.

The first spacesuits developed in the late 1950s were designed mainly to protect astronauts in case their capsule sprang a leak and all the air escaped. On 18 March 1965 Soviet cosmonaut Alexei Leonov became the first man to 'walk' in space. He squeezed into the cramped flexible airlock of his Voskhod capsule and pushed himself outside. For ten minutes he enjoyed the exhilarating sensation of spacewalking and then began to pull himself back into the ship, only to discover that his suit, at full pressure, had ballooned outwards so that he could no longer fit into the airlock. Dangerously exhausted by his efforts, Leonov had to let some of the air out of his suit to collapse it so that he could squeeze himself back into the spacecraft.

When Edward White made the first American spacewalk outside a Gemini capsule three months later, the NASA designers of his suit quickly realised that they had made the same mistake as the Soviets – or at least, they would have had they been swapping notes with Soviet scientists at the time. Anyway, White's suit also swelled up like a balloon and he too found it hard to move his arms and legs and had difficulty squeezing back into his seat at the end of his walk. However, at least the Gemini hatch was wide enough to let him back in.

Everyone went away to rethink the spacesuit problem. Today, seen from the outside, a typical spacesuit looks as if it is constructed from stiff yet basically flexible fabric, but this is deceptive. The upper torso and the arms and legs are strengthened with reinforced nylon fabric that doesn't expand (or 'balloon') when inflated with air. The suit maintains a constant interior volume, even when pressurised from the inside with air. Rubberised bellows at the shoulders, elbows, hips and knees allow an astronaut to bend his or her arms and legs without too much difficulty, but they do have to work quite hard against the resistance of the suit's materials. When the outer white fabric layers are stripped away the semi-rigid design of a suit is revealed like some kind of alien robot. Spacesuits are far from perfect. Spacewalkers are frustrated by the simplest things, such as the lack of visibility provided by their helmets. The next generation of suit for the International Space Station will include a larger helmet with all-round visibility.

The pumps and hardware in the backpack are quite complicated, yet there is plenty of scope for you to redesign the more basic parts of the suit, especially the main fabric covering. You don't necessarily have to seek multi-million dollar funding. All you need are some good ideas and a sewing machine.

bruised fingers

Gloves are perhaps the most serious issue. They are too stiff to handle delicate tools, yet strong fabric around the fingers is essential for safety, because gloves are more at risk of puncturing through wear and tear than any other part of a suit. As your colleagues help you peel off your heavy suit after a long session, they will scarcely bother to notice your sweaty aroma – or your bruised fingers. Spacesuits are incredibly uncomfortable and working through heavy gloves can leave your fingers completely numb, so you don't notice the severe chafing. In sunlight the suits' cooling system works over-time to protect you from being broiled alive. In the shade, temperatures sink to minus 125° Fahrenheit (–87.2° Celsius). Sweat clings to the skin, becoming cold and clammy as the temperature sinks. With sunrise and sunset alternating every 90 minutes as your ship speeds around the earth, a long spacewalk can throw a suit's cooling circuits into confusion. In February 1995 astronauts Michael Foale and Bernard Harris had to be pulled in early from a spacewalk in case they lost their fingers to frostbite. Kathy Thornton, a veteran of the famous Hubble repair missions, says that working in a space-suit is 'like trying to fix the carburettor on your truck while wearing thick baseball mittens, and it's like you're skidding about on an ice-rink because you've got no purchase. On TV it looks a breeze, but it's truly exhausting.'

In response to this, Peter Homer, an engineer from Southwest Harbour, Maine, sat at his dining-room table one day and stitched together a mock-up for a new and improved spacesuit glove. It didn't matter that his materials were somewhat less sophisticated than NASA's or that his home-made product fell far short of anything required for a real space mission. His angle of attack was simply to focus on cutting out the bruising and chafing. Homer thought anew about the stitching pattern and how the interior layers of the glove rub up against the outer layers. His design, he argued, could easily be incorporated into real gloves made from more exotic materials. NASA was impressed, awarding him $200,000 from its Centennial Prize fund. Homer has no plans to build a vehicle or travel into orbit himself, yet he has become a *player* in the space business.

Creating a comfortable, lightweight suit for the future remains a challenge and NASA is open to ideas from any quarter. Explorers on Mars will need a suit far tougher than any previous design, but at the same time it'll have to be comfortable, because they'll be wearing them day after day; and while the Apollo suits were built to survive lunar-surface missions lasting at most two or three days, Martian suits will need to function for weeks or months on end without springing a leak. What's more (yes, the problems just keep piling up), the orange-red dust that covers almost the entire surface of Mars is chemically corrosive. Every time you climb in or out of your suit there's a danger that your helmet's neck-ring, glove joints and oxygen connectors will become clogged with dust. This might be avoided by keeping the suit in one piece at all times. The life-support backpack could be hinged like a doorway for the wearer to climb in and out through. This would reduce the number of places where dust might become trapped.

The exterior of the suit could also be constructed from a smooth, hard shell that can be wiped clean of dust as easily as wiping down the paintwork of a car. A rigid exterior should also provide extra protection against punctures. But then your problem becomes how to move inside such a lumbering monster outfit? All of these requirements threaten to make suits for Mars very heavy. The Apollo moon suit was designed to operate in lunar gravity, which is 1/6th that of the earth. On Mars, that suit would weigh almost twice as much and its wearer would soon become exhausted by the effort of moving. Martian suits will have to be made from very light, strong materials, perhaps not yet invented.

It could be that the traditional bulkiness of spacesuit designs has been an evolutionary mistake. There's no particular reason why a constant-body spacesuit shouldn't be almost exactly the same shape as the human body. Skin-tight materials could make a spacesuit almost as sleek and stylish as a ski racer's outfit. If the layer of air between the wearer's skin and the inner layer of the suit fabric is kept as thin as possible, then you won't get the 'ballooning' effect that dogged early astronauts. Anyway, questions such as these are worth pondering. If you can come up with the right fabric and the right stitching, you can leave the backpack and helmet complications to the bigger companies.

yes, they really did land

Thousands of conspiracy theorists around the world continue to claim that the Apollo 11 astronauts never landed on the moon. They argue that the images we've all seen of Neil Armstrong stepping onto the lunar surface in July 1969 were faked in a giant TV studio in one of the biggest hoaxes of all time. Here's how you politely beg to differ.

The simplest argument that these conspiracists use is that there should only be one very concentrated source of light in the Apollo lunar pictures – the sun – but many of the shots look as if they were snapped using some kind of much broader and softer illumination. The harsh sunlight on the moon should create deep black shadows, with no subtle shading in between. Yet in most of the shots the astronauts seem surprisingly softly lit, like male models showing off the latest craze in extremely clumsy clothing. The shadows look as if they have been filled in by lights other than the Sun. The explanation is simple. The astronauts and their lunar module are standing on a gigantic grey-white photographic reflector. The moon's surface is a bright grey-white, as any pair of lovers will notice when they look up and see how brightly the moon bounces sunlight back into space. Photographers and movie-makers among you will know all about using big reflector panels to 'fill in' the shadow areas of your subject, particularly when counterbalancing bright directional sunlight. On the moon the surface terrain works a treat and the astronauts are perfectly lit.

The other argument from the conspiracy addicts is that countless stars should be visible in the black and airless lunar sky, yet there's not so much as a single speck. This is a mystery that any amateur photographer can solve after a moment's thought, but the way you explain it will make you seem like a true expert. An astronaut in a white spacesuit in bright sunlight needs an exposure of only a few fractions of a second, but stars are too distant and dim to register on normal photographic film, unless exposure times of several seconds are used. That's also why shuttles and space stations photographed in orbit invariably appear against the same pure black sky. Time and again, space travellers tell of the vast swathes of stars they can see, yet those stars never appear in the photos.

NASA's astronauts and their Soviet rivals accepted great personal risks in the struggle to reach the moon and in January 1967 three of them, Gus Grissom, Ed White and Roger Chaffee, died in the most horrific circumstances imaginable, sealed inside a cramped capsule while it burst into flames. A few months later a Russian pilot, Vladimir Komarov, smashed to earth when his parachutes failed and his capsule thudded into the Russian steppes like an unrestrained meteorite. Then, in 1970, Jim Lovell, Fred Haise and Jack Swigert nearly lost their lives when their main oxygen tank exploded on the outward trip to the moon. You might point out to the conspiracy theorists that it's pretty mean-minded of them to claim (as some do) that the famously dangerous Apollo 13 mission never left the safety of a film studio.

seven   the red planet – and beyond

When the Italian astronomer Giovanni Schiaparelli drew the first map of Mars in 1877 he had nothing more to guide him than the blurred images captured by his simple telescope. He recorded, as best he could, several large dark plains loosely connected by much narrower features, which he labeled *canali*. In Italian, this simply means a groove or channel. An enduring myth was created when Schiaparelli's work was translated for English-reading astronomers. The word *canali* was interpreted as 'canals', artificial water-bearing structures created by an intelligent civilisation. In 1894 Percival

Lowell, a brilliant but misguided American astronomer, examined Mars from a high-altitude observatory in Flagstaff, Arizona. He was sure he could see those canals crisscrossing the entire Martian surface. In 1896 he published his findings: 'On Mars we see the products of an intelligence. There is a network of irrigation. Certainly we see hints of beings in advance of us.'

What might Lowell have 'seen' if he had never encountered those flawed translations of Schiaparelli's findings? We will never know. Lowell's work was mocked by other astronomers, but his misapprehensions turned out to be far more entertaining than the reality. His proposal that an ancient water-starved civilisation had girdled an entire world with canals and pumping stations was too good to waste. Newspapers and popular writers took up Lowell's ideas with enthusiasm, not least H. G. Wells in his classic novel *The War of the Worlds* (1896), the tale of an invasion of earth undertaken by intelligent but merciless Martians. The novel's opening lines retain all their power today:

No one would have believed in the last years of the nineteenth century that this world was being watched keenly and closely by intelligences greater than man's and yet as mortal as his own . . . Across the gulf of space, minds that are to our minds as ours are to the beasts that perish, intellects vast and cool and unsympathetic, regarded this earth with envious eyes, and slowly and surely drew their plans against us.

By the end of the nineteenth century astronomers knew better. Spectral analysis of sunlight reflected from the Martian surface or glancing obliquely through its atmosphere provided some sobering data. The air is whisper-thin, incredibly cold and comprised mainly of carbon dioxide. Neither Lowell nor the fiction writers allowed these observations to interfere with their grand visions of an intelligent civilisation. The dominant twentieth-century image of Mars was promoted in pulp-fiction magazines. Edgar Rice Burroughs, creator of the jungle hero Tarzan, wrote a series of fantasies set on Mars, which is called Barsoom by its inhabitants. Readers around the world developed a taste for life on Barsoom. Canals were the least of its attractions. It also boasted beautiful princesses . . . Mars hit the newspaper headlines in 1938 when Howard Koch and Orson Welles turned *The War of the Worlds* into a radio play that included a realistic-sounding news broadcast. Many Americans thought the bulletin was genuine and that Martian invaders had really landed in New Jersey. In 2005 the film maker Steven Spielberg made a spectacular and unsettlingly fatalistic version of *The War of the Worlds* aimed at a post-9/11 audience.

Science fiction has its uses on earth, but real-life discoveries aren't always so flexible. During the 1950s most scientists reconciled themselves to the fact that Mars simply isn't the kind of place where intelligent life hangs out. However, they still expected *something*. As the space planners in America and Russia developed their new and potent rocket technologies, Mars became a favoured target. A human mission seemed possible, in theory at least, for the next generation. In *Das Marsprojekt* (1952) Wernher von Braun outlined how 1950s technology might achieve a mission. Ten ships, 4,000 tons apiece, would make the trip. It seems poignant, now, that anyone could ever have imagined launching such an impressive space armada.

Arthur C. Clarke wrote a fine science-fiction novel *The Sands of Mars* (1951), depicting a self-sufficient colony of earthlings breaking away from their mother planet, while Ray Bradbury used the planet as the setting for one of the most famous science-fiction works of the twentieth century, *The Martian Chronicles* (1950), in which Mars was inhabited by the ghostly remnants of an alien culture. Bradbury also depicted the carelessness of humans when they enter a new and unknown environment. Robert Heinlein promoted his hopes for the betterment of mankind by using Martian intelligence as a model in his novel *Stranger in a Strange Land* (1961). Its influence fed into the 1960s counterculture.

So when did our current perception of Mars first take hold of our imaginations: the dusty, desolate, boulder-strewn world we *think* we know today? By the 1960s those canals had been firmly ruled out by the first robotic space surveys. However, in a report called *Conquering the Sun's Empire* prepared by NASA as recently as 1963, the authors stated, referring to Mars: 'We can feel reasonably confident that primitive life exists.' A decade later, a pair of NASA robot landers, Viking I and II, sampled the Martian surface soils using instruments designed to detect life. The results were inconclusive, but for a while it seemed as if the long-held dream of finding even the buggiest bugs on the Red Planet had withered in the cold, thin atmosphere and dry soils of a lifeless world.

mars@click.mouse

There was a time, just four or five decades ago, when the planets in our solar system were almost unimaginably distant and strange to us. It seemed a great achievement if we could send simple space probes to obtain even a brief glimpse of these worlds. Entire missions were planned and launched whose purpose, after many months' hurtling across the void, would be to arrive in the vicinity of their target planets, then skim past for just a few precious hours while snapping a couple of dozen fragile images, before once again skittering off into the wastes of space. In the early years of rocket exploration it wasn't always possible to slow down a craft and put it into orbit around such distant worlds as Mars or Venus. It was enough of a challenge simply to get them into the general vicinity.

On 15 July 1965 Mariner IV flew past Mars after a trouble-free journey of 230 days, coming within about 6,000 miles of the planet at its closest approach. In these early days of remote-imaging technology the transmission of pictures from deep space was no simple matter. Mariner's camera recorded 22 images onto a strip of photographic negative film, which was then processed internally on a convoluted series of rollers. A scanning device, similar in principle to an old-fashioned fax machine, then translated the results into radio signals for transmission to the waiting earth, with pulses corresponding to light or dark areas on the negatives. Each picture took eight hours to transmit. In modern computer terms this would represent a snail-like rate of eight bits per second. In order to send its entire batch of 22 images, Mariner's radio gear beamed away solidly for ten days, even as the spacecraft itself left Mars far behind and hurtled into deepest space. The original film's negative black-and-white values were then electronically reversed by the earth-based receiving equipment to produce positive images: the first close-up look at Mars in fuzzy frames of 200 scan lines, each consisting of a string of 200 tiny dots.

Today those images seem crude, but at the same time deeply touching. So much effort was invested by so many people to achieve what we would now regard as so little reward. The merest fuzzy rumour of an image from deep space was once considered a technological miracle. Today we routinely expect high-resolution data from the farthest reaches of the solar system, even from distant galaxies, via our space telescopes. Although we benefit from this perfection, some of the ghostly mystery of space has perhaps been stripped away by the searing clarity of our vision.

armchair astronomy

The 1997 Mars Pathfinder mission opened a new era of Internet space exploration. As the little wheeled rover *Sojourner* beamed back its startlingly clear images of the planet's surface, the relevant NASA websites received a world-record 33 million 'hits' on 4 July, touchdown day. Four days later that figure had risen to a total of 47 million. A small Internet team (led by Kirk Goodall of the Jet Propulsion Laboratory in California) had revolutionised public awareness of planetary exploration on a budget so small it made almost no impact on Pathfinder's overall costs. We now expect to be able to log on and retrieve data from planetary rovers or deep space probes without a second thought. The visionaries of half a century ago imagined that we would visit the farthest reaches of the cosmos. What they didn't expect was that those regions would be brought to us, without our so much as having to leave the house.

As Arthur C. Clarke has written: 'One day we will not travel in spaceships. We will *be* spaceships.' In a sense we already are. Robot probes are extensions of our minds and our physical reach, allowing us to scrabble in the soil of far-distant worlds with remotely operated claws. Today's technology has made Mars seem almost as close and accessible as our own backyard. High-definition images of its surface can be downloaded on home computers within a few days or even hours of transmission. Ground controllers need time to process the images, but essentially they are made available to the public as soon as they are ready. We take for granted the God's-eye views we now have of Mars, Jupiter, Saturn, dozens of moons and even several major asteroids.

Does that mean, then, that we no longer have to go to all the trouble of turning up in person to hold alien dust in our own hands? This has been the great debate ever since the Space Age began. Logic suggests that machine probes are the safest and most cost-efficient tools for space exploration. Instinct and emotion cause many of us to think differently. The world's major space agencies are gearing up, now, for a crack at the Red Planet in the coming generation, using astronauts as well as machines. If you are reading this book and you've celebrated your fortieth birthday already, then your chances of getting involved in these long-term quests are modest, although by no means hopeless. On the other hand, if I'm talking to any young folk out there, then pay attention, because Mars might become your career one day.

keeping it simple

Forty years ago a senior Apollo manager called Joe Shea was listening to a presentation from one of NASA's contractors. They were wondering how to protect the spacecraft against the fearsome heat of the sun on one side and the freezing cold of space on the shadowed side. 'They came to me and said they didn't know how to do it,' recalled Shea. 'I asked, them, "How much extra money do you need to work it out?" And they said, "At least $60 million." So I thought about it, and I said, "Well, *why* is this the requirement?"' It turned out that the capsule's resin heat shield, although perfect for the searing temperatures of re-entry, was not so happy when cold. The shadowed side of the spacecraft would be so cold, and the sunlit side so warm that the heat shield would shatter like a delicate china plate taken out of the fridge and dipped in hot water. In a flash, Shea saw the answer: 'I came up with what I called the "rotisserie mode". I said, "Let's just spin the whole spacecraft very slowly, once a minute or less."' That meant the sun's radiation would be spread evenly across the entire exterior of the ship and nothing would ever become too hot or too cold. The entire problem was solved in the simplest way possible – and at no additional cost. Today, some engineers are wondering if NASA can apply that kind of thinking to simplify a human mission to Mars.

But you know what? NASA is still not very good at thinking outside the box. Still reeling from the criticism it endured after the *Columbia* shuttle disaster, the lumbering space agency is slowly but surely opening itself up to clever ideas from the outside world. Maybe you can figure out what NASA can't . . . For instance, until recently most Mars mission plans have been so complicated and costly they've failed to win wide political support. At the Martin Marietta Corporation in the 1980s, engineer Robert Zubrin was appalled by the costs NASA had in mind. It seemed to him that building giant spaceships had become more important than the mission itself. 'I fired off a memo saying it isn't enough simply to reach your destination. You have to do something useful when you get there. I thought the existing plans were totally wrong and too expensive, and many people at NASA were upset when I spoke out of turn.' Industry didn't appreciate Zubrin either. 'Aerospace companies usually tell NASA exactly what they want to hear, because that's the way to make a sale. I was proposing to tell the truth, whether NASA liked it or not. Theirs was the *worst* and most inefficient way to get to Mars.'

Al Schallenmuller, Marietta's chief of civilian space systems, fondly remembered working as an engineer on NASA's robotic Viking lander missions of 1976, and he was keen to see a human Mars mission if it were at all possible. He allowed Zubrin and a dozen colleagues some time to rewrite the company's official sales pitch to NASA. By February 1990 the team had reduced the Space Station's role in the Mars mission, cut the weight of the outbound ship in half and slashed the costs. But Zubrin still wasn't satisfied. Time spent on the Martian surface was still only four weeks out of an 18-month round-trip. The earth orbital construction of the ship annoyed him, because it cost money and wasted fuel; and the ship was still much heavier than he wanted. He came up with 'Mars Direct', a plan that cut size, weight and costs to the bone, required no time in earth orbit and eliminated the Space Station's role as a construction base and stopping-off point. Mars Direct is so neat and compact, it just might be fundable by a private industry such as yours . . .

'Most Mars plans call for a huge mother ship to circle the planet and send down small landing teams, which then come up, rendezvous with the mother ship again and fly home,' says Zubrin. 'I call it the *Battlestar Galactica* approach. Why have the mother ship at all, hanging around in orbit and doing nothing? In Mars Direct, you fly small pieces of hardware directly to the Martian surface, and then a small Earth Return Vehicle (ERV) fires you off the surface, and you head back home.' This smaller style of ship would mean staying confined in relatively cramped cabins for the six-month outward and eight-month return trips, but as Zubrin points out, 'We know from the Mir and International Space Station experience that you can tolerate that if you're sufficiently motivated. You don't have to build giant space cruisers to go to Mars.'

A huge amount of fuel (and therefore spacecraft size and weight) is saved by launching only when earth and Mars swing close to each other in their orbits and are both on the same side of the sun at once. This happens roughly every two years, so Mars Direct employs a rolling schedule of missions to coincide with these close planetary approaches. The downside to this 'low-energy' scheme is that your mission takes well over two years from start to finish, because you have to wait 18 months on Mars until the planet swings close by the earth again before you can set off on the return trip. On the face of it this lengthens your exposure to cosmic radiation hazards. But while Mars Direct involves a longer stay on the planet than NASA originally had in mind, you'll actually spend a little less time in deep space, where the radiation hazard is greatest. Supporters of Mars Direct also point out that space radiation doesn't kill you. As we've discussed earlier, what it actually does is increase your risk of radiation-related cancers in later life. So far, the death rate of astronauts shows that (accidents aside) their longevity is no different from normal. 'We've spent more than 40 years sending astronauts to space stations for long periods,' Zubrin argues. 'There is nothing more to learn about radiation that we don't know already.'

However, weightlessness is another problem altogether. We know that astronauts on extended space missions lose body mass, muscle tone and bone strength. Your earth-orbiting space hotel might feature weightlessness as part of its appeal, but a long Mars mission is different. You may want to think about artificial gravity to keep your crew strong and healthy. A long tether stretching between two modules can create a basic version of the gently rotating artificial-gravity space stations so favoured by rocket visionaries in the 1950s; only instead of a spinning wheel, your ship will look more like a tumbling stick. You will arrive at Mars in pretty good shape, yet without having to travel in some vast and complicated starship. For the homeward voyage only one module (the ERV) is employed, so the tether and its gift of artificial gravity will be lost. If, as a result, you're a little less than superstrong on your return, it doesn't matter so much. There will be a reception team and medical support on standby. You should land back on earth in no worse shape than astronauts coming home from space stations.

living off the land

There is always a chance that you might get stranded on Mars. What if your lander crashes or fails to take off at the end of your stay? Again, this is a simple problem to fix. Uncrewed Earth Return Vehicles (ERVs) are launched to Mars under computer control. The modules then touch down and report back their status. Only when they are sitting safe on the surface, ready to lift off for home, do you set off from earth to join them. Those ERV modules will land with their fuel tanks half empty, thus saving even more weight and reducing the cost of launching them from earth. The key to this saving is that Mars's atmosphere consists mainly of carbon dioxide. An In-Situ Resource Utilisation (ISRU) plant can pump this through a nickel catalyst, adding a trace of hydrogen into the mixing chamber. The catalyst splits the carbon dioxide, liberates the oxygen and combines it with the hydrogen to make water. The freed carbon reacts with spare hydrogen to create methane – rocket fuel for your return trip. Six tons of hydrogen carried to Mars could be converted into 108 tons of methane and oxygen.

No new technology has to be invented to make ISRU work. One by one, Zubrin and his army of supporters in a highly organised lobbying group are trying to knock down the anxieties and cost implications that, so far, have prevented humans from flying to Mars. There will be no easy answers and there are bound to be some unpleasant surprises as well as sweet victories. There is no guaranteed formula for success. There's only a varying degree of risk to be played off against time, cost and political will.

Your best bet is to get the Russians on board. American technicians preparing for the 1975 Apollo-Soyuz docking mission were alarmed by the Soyuz spacecraft's primitive computer. NASA had a stack of printouts in which complex orbital mechanics were transcribed into digital code. How was all that data supposed to fit inside the Russians' crude circuitry? The Russians shrugged and presented a few simple equations on a single sheet of paper. The shame-faced Americans saw how the elegant Russian mathematics could work inside even the simplest computers.

Look at Russia's achievements. It was the first country to launch a satellite, Sputnik, in October 1957. Then came the first living payload, the dog Laika; then the first man, Yuri Gagarin; the first spacewalk by Alexei Leonov; the first woman into space, Valentina Tereshkova; and the first robotic lunar rover, Lunokhod ('moon walker'). Then came the first space stations, and the first robotic docking of modules. So let us celebrate the fact that in 2007 NASA signed a $20 million deal with the Russian Space Agency for one of the most essential technologies of the future (and yet another Russian first), a space toilet capable of extracting drinkable water from astronauts' urine. Without such a device, all hopes of a human mission to Mars are doomed to failure.

mars right here

If you ever happen to visit Devon Island in the high arctic Nunavut Territory of Northern Canada, you might be surprised by what you find. There's a dramatic 30-mile wide crater, gouged out 23 million years ago by a meteorite. The harsh rocky landscape looks like the surface of another planet, and when you spot a team of spacesuited explorers picking up rock samples inside the crater, you might think you really are on an alien world. In 1997 NASA scientist Pascal Lee set up a project to explore Devon Island's impact crater. When he saw it for the first time, he knew he'd found a perfect place for training Mars explorers. So he contacted the Mars Society, a 5,000-strong group of scientists, engineers and space enthusiasts dedicated to putting humans on the Red Planet as soon as possible. Armed with more than a million dollars worth of sponsorship from the Discovery TV channel and the Flashline software company, the Mars Society constructed a replica Mars Habitat (or 'Hab') on Devon Island. It's freezing cold here, but also very dry. As Lee says, 'It's as close to Mars as we can get without leaving earth.'

With support from NASA, the Mars Society borrowed a US Marines C-130 transport plane to fly all their equipment to the island. There was nowhere in the rough landscape for the huge aircraft to touch down. Instead, everything was packed into crates and dropped by parachute. One important load slipped out of its harness in mid-air and was smashed to pieces when it hit the ground. The construction team lost the special crane they needed to lift the Hab's walls into position. Thousands of kilometres from help, and in bitterly cold conditions, they had to improvise a way of finishing the Hab. It was a dramatic rehearsal for the kind of emergency that might easily occur during a real Mars mission.

Today the Flashline Mars Arctic Research Station (FMARS) is one of the strangest games of make-believe anyone ever played. Teams of scientists are learning how to live and work on the Red Planet. You can volunteer, but be aware that the rules are very strict. You don't go outside the Hab except in a spacesuit. You even have to wait in the airlock for 30 minutes while the pressures adjust. (The atmosphere on Mars is less than 2 per cent as dense as the earth's.) Then, when you come back into the Hab, you'll have to vacuum-clean your suit to get rid of all the dust from the outside. Martian dust is corrosive and dangerous to breathe, and the suits' moving parts have to be kept clean so they can be reused day after day. (And it'll be no good saying you'll do it later. You have to be disciplined.)

OK, the Mars Society spacesuits are just costumes, the Devon Island soils are perfectly safe and the atmosphere outside the Hab is really just the same as on the inside. But everyone taking part plays this game as if it were real. They are finding out how to live on Mars, how to get along with each other in a cramped module, and how to work efficiently on the Martian surface. It's better to practise all this stuff on earth so that future explorers won't make deadly mistakes on Mars. Lee compares the Hab experience to army training. 'It's basically like a military field exercise, only instead of being about war, it's about exploration. Mars will be a challenging place, so you have to work out your tactics before you go on the real mission.'

The Hab crew communicates by radio with NASA's Ames Research Laboratory in California. But no one is allowed a normal phone-style conversation, because in a real Mars mission radio signals take, on average, ten minutes to travel to earth. That means you have to wait another ten minutes before you get a reply. A time delay is built into the Hab's radio link. It can seem like ages when you want urgent advice, but it's important for everything to be as realistic as possible. A second Hab has been built in the rocky canyon lands of southern Utah. It's much warmer here than on Mars, but the landscape looks amazingly similar. Hab scientists are testing robot rovers, prototypes of semi-intelligent machines that will one day work alongside real astronauts, gathering rocks and drilling for soil samples.

So what's the most important thing that Pascal Lee learned from the Hab experience? 'When things are going against you, what you really need on your crew are people who know how to laugh. On a real mission to Mars, if you lose your sense of humour, you're finished.'

asteroid alert

Apart from going to Mars, one of the projects you might consider is digging mineral wealth out of tumbling chunks of rock, while at the same time saving humanity from total destruction. In the lonely hinterlands of space between the orbits of Mars and Jupiter lies the asteroid belt, a 100 million-mile-wide band where countless chunks of rubble circle the sun. These random rocks are left over from the formation of the solar system's planets four-and-then-some billion years ago. The gravitational influences of Jupiter and Mars keep most asteroids within the belt. Jupiter even hoovers up many of the wanderers, sucking them into its gaseous depths. But every so often the gravitational forces run amok and stray asteroids are thrown out of the belt. Like cars skidding out of control on a busy motorway, they drift towards the sun and across the paths of oncoming planets. So you have to expect the occasional pile-up.

On 14 January 2000 the Near Earth Asteroid Rendezvous (NEAR) spacecraft achieved our first close encounter with Asteroid 433, also known as 'Eros'. In a complex four-year journey, NEAR had travelled past Mars and then fallen back towards the earth during 1998 to pick up extra speed. When it reached Eros it was travelling round the sun at almost exactly the same speed as its target. Eros is 20 miles long and about eight wide, shaped like a peanut and just massive enough to produce a gravitational field. (If you stood on the surface, you'd weigh about as much as a foil-wrapped bag of nuts.) Mission controllers had to work out, backwards as it were, how this feeble force would influence NEAR as it approached, because that was the only way of calculating a stable orbit. The tiny probe had to circle Eros at less than 10 mph in order not to drift away.

NEAR's instruments tell us that Eros has a rocky crust almost as hard and dense as that of our own planet. Its internal structure is not a loose clump of boulders and dust, as some theories had predicted. It's a single massive chunk, albeit deeply fractured after a long history of collisions with other asteroids. However there *are* multiple elements to Eros. Approximately a million boulders litter the surface, each the size of a house or larger. If this is typical for an asteroid, then it's bad news. Any number of asteroids approaching the earth could have a similar collection of boulders, all as dangerous as multiple warheads on a missile. Pictures reveal a staggering 100,000 craters, ranging in diameter from a few feet to as much as 3 miles across. In some places radar scans show impact debris in the form of dust and rubble ('regolith') lying in troughs 300 feet deep. At least a thousand similarly sized asteroids regularly sweep across the earth's path and there are countless smaller ones out there, too, which are impossible to track. If a rock the size of a Greyhound bus hit New York, the city would be destroyed. Anything bigger than a mile across could wipe out human civilisation.

In August 2004 a Near Earth Object (NEO) task force convened by science minister Lord Sainsbury reported to Tony Blair's government that asteroids should be regarded as a threat to the security of Britain. Whenever the lives of more than 10,000 British citizens are endangered, this is called an 'intolerable' figure and official guidelines say that an emergency should be declared. The troops should be ready at a moment's notice and food stockpiled. Power-plant explosions and nuclear war are hazards that the Government already takes seriously, but storms, earthquakes, plagues and other potential bad days are also on their list. If the chances of catastrophe are greater than 1 in 100,000 in any given year, emergency plans have to be considered, even when there is no immediate threat. Lord Sainsbury's team showed that the chances of an asteroid collision fall well within this range.

Most asteroids are too small to spot, except by accident, when a thin streak of light smudges across a photographic plate while a telescope is searching the sky on other routine business. The reason why asteroids are so rarely tracked is that astronomers have to arrange their telescope access months, sometimes even years, in advance. Committees judge their applications and astronomers usually choose targets like stars and galaxies, because they already know where to find them. Even the fluke asteroid discoveries are often lost a few months after they've been identified.

According to the best geological evidence, half-mile-sized asteroids have slammed into the earth about once in every 100,000 years. That's exactly the minimum annual risk factor that activates the Government's guidelines. Smaller asteroids probably hit the earth once every 10,000 years. The planet as a whole easily survives these small disasters and it's improbable that *your* city in particular would sustain a direct hit, because it's such a small target in comparison to the earth as a whole. But if it doesn't get you in the first shot, it might well catch up with you later. A piddling little 500-foot diameter asteroid splashing into in the Atlantic Ocean at 10 miles per second could easily drown our coastal cities under a tidal wave, killing thousands, even millions, of people. OK, so the risk, considered on a day-by-day basis hardly gives us any reason to panic, but a one-off impact any time in the future would cause so much damage that it's worth thinking about how to avoid it.

dealing with them

One solution would be to classify asteroids as a hostile threat, a military problem requiring funds from the defence establishment. That's not a prospect that pleases everybody. It would give the US aerospace companies all the excuses they need to develop their controversial 'Star Wars' weapons in space. The military would like nothing better than to develop nuclear rockets capable of 'taking out' a menacing asteroid, but the chances are this would create more problems than it solved.

Blowing up an asteroid would create a swarm of debris almost as dangerous as the original big lump. A more subtle approach would be to create a nuclear burst alongside an asteroid, deflecting it from its course rather than smashing it into pieces. But we don't have nuclear missiles capable of reaching deep enough into space to catch up with an asteroid; and is it really a good idea to start building them? Even if we did, the missiles' journey time might still be two years and more. We would need three or four year's advance warning of an asteroid hazard, plus another year for the global community to discuss what to do and who should be in charge. At the moment, placing any kind of nuclear warhead above the earth's atmosphere is forbidden by international law.

Perhaps the best way to deal with asteroids would be to leave the nuclear weapons in their silos. Instead, we could send simple robot probes to push an asteroid gently off its doom-laden course. You could trap sunlight and turn it into electricity to power an ion gun, exerting a very small force on the asteroid, but over a long time. This would just nudge it out of its current orbit. The tiniest thrust would make all the difference, and this delicate technique would also prevent lethal swarms of debris flying off the asteroid's surface. The downside of this otherwise straightforward technique is that you would need perhaps a decade of advance warning. Your probe's long flight time is only half the problem. You would have to deflect the asteroid while it was immensely far from actually hitting the earth, so ion engines couldn't cope with any sudden emergencies, because their effect on the asteroid would be so weak. For now, the best thing you can do is to learn more about what the earth is up against. Either that, or start thinking about the mineral rights.

The famous astronomer and astrochemist Carl Sagan predicted that the skills needed to deflect a dangerous asteroid away from earth might just as easily be used to steer a harmless one towards us. Rogue nations or space-based terrorists could blackmail the entire planet. In 1994 he warned, 'I keep hearing that only a madman would do something like that. I have to remind myself that madmen really exist.' Failing that, orbiting entrepreneurs could destroy us by accident, seeking riches but harvesting destruction instead. Some asteroids are rich in platinum, one of the most precious metals; and others consist almost entirely of pure metals. Every scoopful of drilled material would yield a profit if only the world's mining companies could get hold of an asteroid in the first place.

It will be a while yet before we can send human prospectors into the asteroid belt and the cost of that operation probably wouldn't be justified. By sending robot probes to do the legwork, you could steer asteroids in our direction just as easily as keeping them away from us. In a few years they would arrive in low earth orbit, easily within reach of astronaut miners. There's even a scheme to 'aerobrake' incoming asteroids in the same way that probes are routinely skimmed through the atmosphere of Mars to slow them down. But the slightest miscalculation could be disastrous. It would be a terrible irony if the next major asteroid to hit us was brought here deliberately. You might end up with a massive compensation claim.

going nuclear

In January 2003, with asteroids and other distant targets in mind, the White House announced its support for a new NASA initiative called Project Prometheus. It was supposed to revolutionise space exploration, enabling probes to reach the most distant planets with ease. It all sounded intriguing at the time, but less than a week after the announcement, the space shuttle *Columbia* was destroyed during re-entry. As attention turned to fixing NASA's most immediate problems, the long-term dream of nuclear rocketry slipped off the agenda and Prometheus was quietly shoved onto the back-burner.

At least, it was as far as the Press were concerned. Behind the scenes, NASA forged ahead with plans for one of the most radical rethinks of space technology in a generation. Prometheus is the family name for a new brand of atomic-powered interplanetary probes. Miniature plutonium reactors provide electricity for ion drive thrusters, which shoot out streams of gas particles and can run continuously for years, building up colossal velocities. On top of that, 100kW of electricity is generated as a by-product of the nuclear reactions to energise the science instruments, computers and radio transmitters. Data streams back to earth at very high speeds: 10 megabytes per second, as opposed to the mere 100 kilobytes per second limping out of today's probes.

The difference in power between Prometheus missions and anything that has gone before is like comparing candles to floodlights. And that's just the modestly sized robot probes. NASA hopes that a powerful nuclear-fission rocket expelling superheated hydrogen fuel will one day blaze a trail to Mars with the first human explorers riding up front. The flight could take as little as twelve weeks.

NASA would like atomic energy to be the normal power source for spaceships in the next generation. There's just one small problem. Before any nuclear machine can cross the abyss between earth and the other planets it first has to get off the ground. And this is where nuclear power units show their dark side. Conventional rocket boosters will still be needed to get the hardware off the ground and into space. Many concerned scientists and environmental campaigners are worried about the risks. What happens if a carrier rocket explodes in the atmosphere and a nuclear power pack is destroyed? It has happened before. Wayward American and Soviet spy satellites with top-secret plutonium electrical generators have disintegrated in the earth's atmosphere on at least two occasions, in 1964 and 1978. They scattered plutonium dust across vast areas. Scientists are still assessing the damage today.

Since the 1970s low-powered radioisotope thermal generators (RTGs) containing small amounts of plutonium have been routinely used in a wide variety of deep-space probes to generate electricity, because solar panels can't easily power the systems of spacecraft venturing beyond the orbit of Mars, as the sun dwindles to a speck behind them. The famous Pioneer, Voyager and Galileo probes to the outer planets all used RTG technology, and the current generation of wheeled rover robots on Mars survive freezing temperatures for many months with the help of plutonium as well as solar panels.

There are concerns today about the increasing amounts of plutonium being sent into space. The Cassini mission to explore Saturn and its moons was launched in 1997 with more plutonium – 32 kilos – than on any space device ever. NASA reluctantly conceded the possible dangers of an 'inadvertent re-entry'. If Cassini had fallen back into the atmosphere, it would have broken up (it had no heat shield) and five billion people around the world would have received slight radiation exposure. That's not to say that everyone keels over and dies, but plutonium is widely acknowledged as a long-term cancer risk. The statistical chances of people developing cancers in their lifetimes increases markedly if they breathe in even a few atoms of the stuff.

Then there's the *Dr Strangelove* factor. The US military make no secret of their desire to 'weaponise' space, even if they are rather more coy about the exact nature of the hardware they're cooking up. Although NASA insists that nuclear space technology is strictly for the peaceful exploration of the solar system, that pledge can only apply to NASA. Other agencies, such as the US Department of Energy, are closely involved with nuclear projects. New technology developed for space might spread to killer satellites and other power-hungry machines with more sinister purposes than the hunt for scientific knowledge or human adventure in space.

orion's boom

Absolutely *the* weirdest nuclear spaceship was devised in the late 1950s, at a time when, despite the shadow of the bomb, the world half-believed that 'the friendly atom' might yet revolutionise our lives for the better. There would be atomic cars and trains and planes and, of course, atomic spaceships. Project Orion was an engineering study for a spacecraft powered by nuclear-pulse propulsion, an idea first proposed by the Polish mathematician Stanislaw Ulam in 1947. Initiated in 1958, Orion was led by Theodore Taylor at the General Atomics company, with the brilliant British physicist Freeman Dyson as his principal consultant. Both men were convinced that chemical rockets, with their limited payloads and high cost, were the wrong approach to space travel. Orion, they argued, was simple, and above all, affordable. Taylor even proposed that the vehicle be launched from the ground.

The ship was expected to be 16 stories high, 10,000 tons in weight and shaped like the tip of a bullet. The engine consisted of a flat circular 'pusher plate' at the base, with a small hole in its centre, through which miniaturised nuclear bomblets, 'pulse units', were to be ejected at the rate of one every second. Huge shock absorbers between the pusher plate and the main body of the ship evened out the staccato stresses of the bomb blasts and the great monster was supposed to rise into orbit at the rate of one explosion per second. After that, it would head for the moon or Mars at a more modest rate of ten blasts per second. There would be two or three thousand pulse units used per mission . . .

The surviving technical drawings of the proposed ship are grandiose indeed. This was no joke. Orion was expected one day to carry thousand-ton payloads into space: ten times more than NASA's famous Saturn V moon rockets. The project was funded in earnest by the US Air Force and the Defense Advanced Research Projects Agency (DARPA). Taylor and Dyson built several small-scale test vehicles powered by conventional explosive pellets and even launched some mini-Orions a few hundred feet into the air from safe test sites on Point Loma, a three-mile long spit of land guarding the entrance to San Diego harbour and guarded from prying eyes by a stern military presence. The top-secret Orion project wasn't cancelled until 1965, by which time NASA was pursuing a more conventional route to the stars using chemically powered rockets. The age of atomic innocence has long since passed.

some even crazier ideas

Rockets, with their hot gas exhausts, have dominated space flight from the earliest days, but future craft may use less explosive engines to get around the solar system. In fact, one design requires no engines at all. The 'Sunjammer' is pushed along by the constant stream of electrically charged particles generated by our sun (the 'solar wind'). A payload, attached by cables to a huge aluminium foil sail, could cross the solar system without any other power source. Olympic-style sailing races could be organised in space, using robotic devices. Unfortunately, these sails will only work for very lightweight vehicles. Heavy craft like space shuttles, crew transfer pods and Mars exploration ships still need rocket engines to propel and steer them.

And finally, there's the peachiest dream of all space cadets everywhere: are you the budding genius who will one day design a starship? In 1978 the British Interplanetary Society in London drew the plans for a futuristic robot spaceship called *Daedalus*, powered by a nuclear reactor. Its mission is to fly to Barnard's Star, the third nearest star to earth, about six light years away. (A light year is the distance travelled by light in one year: a very long way.) *Daedalus* reaches speeds of 28,000 miles per second, 15 per cent of light-speed, but its journey still takes 50 years. Once it arrives at Barnard's Star its radio signals whizz back to earth at full-on lightspeed (natch), so scientists get their first results 56 years after *Daedalus* sets off. All we have to do now is build the ship – and it's perfectly plausible.

As far as we know today, nothing solid (like a spaceship) can fly at more than a fraction of the speed of light. If aliens wanted to visit us in a ship powered by thruster engines like *Daedalus*, they would have to survive journeys of many years, perhaps even decades or centuries if they came from distant solar systems. Maybe they would go into 'suspended animation' like the human crew in the *Alien* movie. Their journey would seem to pass by in a flash. The only snag is they might return to their home world one day and find that even their great-great grandchildren have long since died of old age.

Maybe aliens wouldn't mind long journeys. In the days when Captain James Cook and his crew explored the Pacific Ocean, a major sea expedition could take several years. At that time, in the 1760s and 1770s, the average lifespan for a sailor was only about 45. He'd have spent most of his life at sea. Aliens might volunteer for long space voyages if their natural lifespans are really long. If they don't have super-advanced technology, they could travel slowly, in a huge space colony the size of a small moon, raising families and passing on command of their ship to future generations.

For faster and more convenient journeys some physicists think it might be possible to create 'wormholes' that instantly connect different regions of space. An advanced ship might be able to navigate through hyperspace, perhaps by harnessing the distorted regions of space-time created by the immense gravity fields of black holes. Or it could warp the space around it with powerful fields of energy to make a wormhole. These ideas are just fantasies at the moment, so don't expect to see warp-driven alien spaceships just yet.

is anybody out there?

In Steven Spielberg's recent movie *War of the Worlds* Hollywood superstar Tom Cruise plays an ordinary man trying to protect his family when hostile invaders arrive from outer space. Could we really be attacked by aliens? Are they actually out there? No one knows for sure, but many scientists believe that we are not alone – and you could play a significant role in settling the debate.

In 1960 a young astronomer named Frank Drake was working with the newly built Greenbank Observatory radio telescope in West Virginia and he thought it might be a good idea to listen for alien signals. So began a project called SETI (Search for Extra-Terrestrial Intelligence), which continues to this day. Drake came up with the

'Greenbank Equation', which looks at all the possible stars in the galaxy (about 400 billion) and focuses on just the tiny fraction that could have planets in stable orbits suitable for life. If there are any creatures on those extremely rare and special worlds, they're most likely to be simple bugs. But the possibilities are still amazing. Even if only one in a billion star systems has exactly the right conditions for intelligent life, there might still be hundreds of advanced civilisations in our galaxy.

But where are they? In the 1940s the physicist Enrico Fermi was working on the secret US Manhattan Project to build the first atomic bomb. In a rare moment of leisure, he and his colleagues chatted about aliens. He asked, 'Why haven't we heard from them?' This simple question became known as Fermi's Paradox. If there are loads of aliens out there, then at least some of them should have sent explorers to earth by now. So far, we've not met any. NASA scientist Chris McKay is leading today's search for microbes on Mars, and he's confident that simple forms of life are widespread in the universe. But he's not so sure about the intelligent ones. He says, 'Why aren't they obvious in all our telescopes? The simplest explanation is that we are the only intelligent species in our galaxy.'

David Grinspoon, a space scientist from the Southwest Research Institute in Colorado, has a different explanation. 'They might not want us to know they are there. They might be protecting us, or protecting themselves from us, and waiting for the right time to make contact.' Aliens may be keeping a low profile, like birdwatchers in a camouflaged hide who don't want to alarm the wildlife.

Until recently, most scientists thought that alien bodies must be different from ours, because their planet won't have the same environment as the earth's. Their biology would be adapted for different conditions, like extreme heat or cold, or a methane atmosphere or perhaps a stronger gravity field. And even if their world is just like ours, they still wouldn't look like us, because of countless tiny variations in their evolutionary history. Just think – suppose our ape ancestors hadn't been so good at survival and modern humans had never evolved. Perhaps meerkats or tigers or lemurs or even *rats* could have become the first creatures on earth to become intelligent! Anything could have happened, so we can't predict what aliens might look like.

Or can we? Today, biologists are studying 'parallel evolution' on earth. Several different families of animal – from fish to mammals to reptiles and insects – have developed some surprisingly similar features. The same number of legs on either side of the body is common in many animals, along with a head and a pair of eyes rather than, say, three eyes or five. Some of the bits on an alien body might be very similar to ours after all. They'd probably have shapes we could more or less recognise; and if they use tools such as spanners and screwdrivers for building their spaceships, then they'd need hands and fingers too.

The most advanced creatures may have replaced their fragile biological bodies with tough, long-lasting machines instead. They might also do just what we already do – send robot probes to explore planets that are too difficult or too far away for them to reach in person. Our first contact with an alien civilisation might be when we stumble across little machines like NASA's wheeled Mars rovers. Only these ones won't have an American flag painted on the side . . .

Even if aliens have bodies just like ours, they surely won't talk in words we can understand. Forget all those TV space creatures conveniently chatting away in English. However, there's one kind of language that scientists are convinced any smart alien would have to share with us and that's mathematics. They wouldn't write their equations in the same symbols, but some concepts would be easy to share with them. The first few 'prime' numbers, for instance (1,2,3,5,7,11,13,17), tapped out with a stick on the side of their spaceship would convince them we were clever too, because they would also know you can't divide those numbers by anything except themselves or 1.

how to find aliens without leaving home

Ever since its foundation in 1960 SETI has used radio telescopes to scan the skies for intelligent signals. Almost everything in the universe creates radio 'noise', because all galaxies and stars release powerful electromagnetic energies. There are mysterious 'gamma-ray' explosions coming from very distant objects far outside our Milky Way. Black holes throw out concentrated beams of X-rays and just about everything in the cosmos radiates radio energies of some kind. SETI's job is to look for signals that are very regular and possibly sent on purpose by alien intelligences. So far nothing conclusive has been found, but some signals have caused a lot of excitement. In 1967 Jocelyn Bell, a young Cambridge radio astronomer, detected a regular and fast-pulsing signal. She and her boss, Anthony Hewish, scribbled a light-hearted note: 'LGM!' Little Green Men. Bell never really believed she'd discovered signs of alien life, but the signals were unusually rapid and absolutely regular, just as if they'd been generated by a machine. Day after day they pulsed at exact intervals of 1.3373 seconds. Since 1967 more than 300 'pulsars' have been discovered within the Milky Way. A pulsar is a small, dense star that spins around its axis in fractions of a second, sending out intense beams of radiation, like a lighthouse gone mad. An intensely powerful gravity field holds the star together.

However, there is one signal that still puzzles SETI astronomers. On the night of 15 August 1977 at the Ohio State University 'Big Ear' radio observatory a strange artificial-seeming transmission was detected, completely unlike any natural cosmic 'noise', and definitely coming from somewhere very far out in space. Volunteer observer Jerry Ehman was monitoring the readings that night, which were printed out as a code on a long roll of paper. He circled the mysterious pulse and wrote 'Wow!' in the margin. Ever since, it's been called the 'Wow!' Signal. Unfortunately, it has never been heard again, so no one is sure what – or who – generated it.

SETI radio telescopes scan thousands of frequencies in the sky at once, then store all the data in computers to be analysed later. Huge amounts of radio noise are recorded and it would be impossible for SETI scientists to search through every little fragment without help. A clever way of speeding things up is to e-mail chunks of radio data to volunteers' ordinary home PCs. When their screensavers come on while they make a cup of coffee a SETI programme does a few minute's sifting through the signals, looking for anything interesting. Today, three million part-time SETI volunteers are helping in the search. You, too, can help look for signs of alien signals.

searching for signs of life

There is a major problem to be solved before we can answer any of our questions about aliens. Space scientists cannot always detect signs of life on the one world where they should be glaringly obvious: the earth. If they can't find it here, they'll have a tough job finding it anywhere else.

When the Galileo space probe swooped by the earth in 1990, building up gravitational momentum for its long trip to Jupiter, all its instruments were pointed towards us for a unique experiment. Strong absorption of light at the red end of the visible spectrum, particularly over the continents, seemed to indicate the presence of chlorophyl, the molecule essential to (mainly green) plant life and photosynthesis. Spectral analysis of sunlight passing through the earth's atmosphere revealed the high oxygen content. Since oxygen is extremely reactive, a dead planet could not support free oxygen in its atmosphere for very long. It takes life (plants, especially) to replenish it constantly, as far as we know. Galileo also spotted small quantities of methane in the atmosphere. Again, as far as we know, methane is usually the product of biology. All these electromagnetically detectable signs of life were clues, not proofs.

In August 2003 a perplexed NASA team reported their failure to sense life in the Atacama Desert of northern Chile. Admittedly, it is one of the world's driest deserts, but it is teeming with life nevertheless. Scientists on the ground were pestered by flies, even as they marvelled at the variety of lichens growing on and under rocks or watched vultures circling overhead: a sure indication that plenty of other animals had to be around somewhere.

But colleagues at the Ames Research Center in California, poring over photos and instrument data transmitted remotely from the field scientists, could detect nothing they considered proof of life. These results were unsettling for scientists trying to develop robot landers and instruments for planetary exploration. If an animal walked in front of a Mars rover, no one but the craziest conspiracy theorist would be left in doubt as to the presence of life, but what if alien entities turned out to be sparser and harder to detect? What subtle data signatures, beamed to us on fragile radio links from an unimaginably remote planet, might prove the existence of life?

This is just one of countless areas of space research where you can get involved, especially if you are just now at school or college. One of the best routes is to train as a scientist, then create experiments and instruments to help settle the riddles of the cosmos. You then put in grant applications for your ideas, applying to the relevant academic and governmental organisations. That way you don't have to spend your own money building the necessary hardware. If you can solve any of the riddles of the universe, a Nobel Prize could be waiting for you a decade or three ahead.

epilogue

Probably one of the most frequent questions you will hear from sceptics will be this: why spend all that money and waste all those resources just messing about in space? Well, the first counter-argument you might deploy is one of relative comparison. 'So, do you take holidays? Oh, yes? Anywhere nice this year? Skiing in Aspen and then kayaking on the Colorado River, because last year you made a special ocean cruise to see the wildlife of the Galapagos and this year you want something more physically adventurous? Sounds wonderful. And how much is that lot costing you then?

Phew! And why do you do all these things? Oh, really? Getting in touch with Nature, appreciating the world around you and learning more about yourself? How interesting! Me too.'

This is the below-the-belt approach, loaded with tactless truths and unlikely to win friends or influence people. A better plan would be for you to persuade your critic of the virtues of space exploration, because, let's face it, if you can't come up with any good reasons for it then your chances of attracting political allies, business sponsors and passenger clients to your project are slim indeed. The space business can be a hard sell. Yours is a pioneering enterprise and therefore most of the rest of humanity is far behind you, curious and slightly fearful and not sure whether to follow in your fire-scorched footsteps. What people do not understand they tend to mock. So sell them on the same dream that impels you beyond the atmosphere. And if you are any kind of proper and decent person that dream will be to see new wonders and then come home and share what you have experienced.

Viewpoint is what human exploration is all about. If you turn the corner at the end of a strange road, your perspective shifts and you see that this neighbourhood is a hilly outcrop of a great city nestling in a valley. So now you understand better the landscape that you are passing through. Similarly, the great (and often vilified) Renaissance astronomers demonstrated that our everyday experience of the sun's arc across the sky is misleading. It is not the sun that moves so much as the earth, rotating on its axis and hurtling, year by year, along its great circular orbit around the sun. We as a species slowly began to understand our cosmic neighbourhood a little better; and even into modern times, the celestial shifts of perspective just keep coming. On Christmas Eve 1968 NASA's Apollo 8 spacecraft fired its braking engine and went into orbit around the moon. If you remember, the lunar-landing module wasn't quite ready to fly and NASA were anxious in case their Soviet competitors pulled a sneaky rabbit out of the hat at the last moment, so the round-the-moon flight was given the green light at short notice.

Apollo 8's commander Frank Borman was a tough character, more concerned with getting the mission right than with its broader poetic significance. At one point he told his crew mates Jim Lovell and Bill Anders to stop daydreaming and focus on their instrument panels. 'I don't want to see you guys looking out the window.' When Lovell accidentally inflated the life vest in his survival harness, Borman glared at him. But when Borman wasn't looking, Lovell and Anders took sneaky glimpses out of a side window at the dwindling earth, 240,000 miles behind them. They were amazed and disturbed that they could block their entire home planet from view just by holding up a thumb.

This mission brought back something far more precious than NASA's scientists could ever have imagined. Colour photographs of the beautiful earth rising above the harsh lunar horizon were printed in every major magazine after the mission and we began to appreciate, perhaps for the first time, just what a tiny, fragile, lonely planet this is, drifting in the infinite black void of space. As Anders observed later, 'We flew all the way to explore the moon, and the most important thing we discovered was the earth.' Critics of the space programme were silenced by the raw emotional value that the Apollo 8 mission delivered, while Borman said how fortunate it was that his crew mates disobeyed his orders and spent as much time as they could snapping pictures out of the capsule's tiny windows. Those snapshots, hastily grabbed pictures that they took *just because they felt like it*, were worth all the risks of Apollo.

In fact one of the great failures of NASA's otherwise magnificent Apollo lunar-landing project was the space agency's excessive zeal for professionalism during every single second of a mission. Senior managers were worried that taxpayers might fret and fume if any of the astronauts took 'time out'. Consequently, the moon walkers' schedules were planned out in advance almost literally to the minute. There were boxes of equipment to unwrap from the sides of the lunar module and a long list of scientific experiments to deploy in a certain order. Then there was x amount of terrain to cover and y pounds of rock samples to be gathered in a given time, and so on. Quite literally, the moon men had to steal the odd moment or two from their busy schedules simply to look around and absorb the impact of what they were doing. No wonder they found it so hard to tell us about it afterwards.

Apollo 8's pictures of earth rising above the lifeless lunar horizon were so new and startling, and so simple in their composition and meaning, we could all understand what they meant. In contrast, the actual moon walks didn't seem to shift our perspective quite so much. We still haven't really been able to share what those Apollo adventurers experienced, because NASA never trained them in that most important of all homecoming tasks: how to convey, in emotional and sensual language, what it was like to visit the moon. Apollo 11's command-module pilot Michael Collins, one of the more thoughtful and literate of the astronauts, said after his epic mission, 'I think a future flight should include a poet, a priest and a philosopher. Then we might get a much better idea of what we saw.' I also have a pet theory that if international diplomatic meetings could be held aboard space stations, then the petty border disputes, trade wrangles, climate arguments and other such life-wasting nonsense might be more easily settled when the politicians become embarrassed at their own parochialism while looking out of the window.

It is true that space exploration is dangerous and difficult. This is another good reason for doing it, because we learn so much by undertaking hard projects and gain so little from repeating easy ones. Why else do we send young people on hazardous adventure-trekking holidays? Once they have grown up and joined the humdrum world of work, we don't expect them to handle office life by setting up tents and breaking open the survival rations, but we *do* expect them to cope better with life in general. Space exploration serves a similar purpose for society as a whole. It keeps us ready for the challenges of the future.

Humans are not angels and the machines of spaceflight are the imperfect products of imperfect people working as best they can within complicated organisations. Yet we reach into the heavens despite our flaws and this is what gives the space adventure its true grandeur. It's not the fact that we explore space fitfully and uneasily, so much as that we manage to do it at all that counts in our favour. Our continued reach into space in the coming generation will continue to be, as John F. Kennedy told us back in 1961, 'difficult' and 'expensive to accomplish'. If the day ever did come when no more bleeping satellites could be found orbiting the earth, except perhaps as long-dead husks – and never again did an elaborate, costly probe swing past the orbit of Mars destined for some far outreach of the solar system – then this would be a clue that human civilisation had essentially failed.

So get out there and keep the adventure going.

acknowledgements

I would like to say thanks to Peter Tallack at Conville and Walsh, who first introduced me to Portobello books. Philip Gwyn Jones, this book's commissioning editor at Portobello, and his colleague Laura Barber, allowed me a great deal of creative leeway, for which I am grateful. Will Whitehorn at the Virgin Galactic company took my calls and gave me valuable insights about the possibilities of space tourism, as did representatives of Burt Rutan's Scaled Composites company. Thanks, as so many times before, to Professor John Logsdon at the Space Policy Institute, George Washington University, for his time and generous access to some often surprisingly frank documents about the political aspects of rocket history. Randy Liebermann, this is a belated thank you for much help given in the past. And finally (but of course I mean first), I would like to thank my family, Fiona, Alma and Oskar, for turning the music down at crucial times, and forcing me to walk the dog even when I claimed to be too busy 'working'.

selected bibliography

Burrough, Bryan, *Dragonfly: NASA and the Crisis Aboard Mir*, Fourth Estate, London, 1999

A fascinating and at times terrifying tale of culture clash and fitful cooperation in orbit between Russian and US space efforts: a good primer if you're planning to make any deals with Russian rocketeers.

Cadbury, Deborah, *Space Race*, Fourth Estate, London, 2005

An adventurous account of the East-West rivalry that kicked off the Space Age in the late 1950s. If we want to plan for where we're going next, it's always good to remember where we've been before.

Chiles, James R., *Inviting Disaster: Lessons From the Edge of Technology*, Collins, London, 2002

A wide-ranging and guiltily exciting analysis of why things fail, and how we misread risks, with plenty of vivid historical examples. You'll want to read this if you're going into the space business.

Gorn, Michael, *NASA: The Complete Illustrated History*, Merrell, New York, 2005

This handsome book gives a balanced 'entry level' account of NASA's successes and occasional failures.

Gray, Mike, *Angle of Attack: Harrison Storms and the Race to the Moon*, W.W. Norton & Co, New York, 2007

Recently reissued, and peppered with charismatic characters, this is an absorbing account of the business deals and political manipulations behind the space manufacturing business.

McDougall, Walter A., *The Heavens and the Earth: A Political History of the Space Age*, Basic Books, 1985

Massive and encyclopaedic 'advanced level' stuff, this Pulitzer prize-winning volume is so beautifully written it might just get you gripped. Perfect bedside reading for space policy wonks.

Murray, Charles & Bly Cox, Catherine, *Apollo*, South Mountain Books, Burkittsville, 2004

Another reissued masterwork, this is a highly accessible account

of the engineering management that made the moon missions possible. Essential reading for anyone planning to revisit that adventure
in the coming decade.

Pendle, George, *Strange Angel: The Otherworldy Life of Rocket
Scientist John Whiteside Parsons*, Weidenfeld & Nicholson, 2005
The intersections between space flight and other forms of human
exploration are the backbone of this startlingly original book.